THE REAL HISTORY OF MAN & EARTH

Myths Magic and Mayhem

Supplement to *Planet X Comets and Earth Changes*

JAMES M. McCANNEY

Α Β Χ Δ Ε Φ Γ Η Ι ϑ Κ Λ Μ Ν Ο Π Θ Ρ Σ Τ Υ Ω Ξ Ψ Ζ

Published By

jmccanneyscience.com press

Minneapolis, Minnesota

1st Printing - 2022

jmccanneyscience.com press
ISBN# 978-0-9722186-3-4

www.jmccsci.com

This book is being released at this time to adjust the direction of researchers who are getting closer to real solutions regarding the history of man and earth BUT who are misguided by incorrect "standard science." Before you can understand the true history, you must have correct physical principles dealing with planetary formation and large comets. Numerous large passing comets (not hitting earth but passing near by) have been the great influencers. The reader must study the McCanney Plasma Discharge Comet Model and all of its ramifications regarding Earth Changes to put the true history of man into perspective.

TABLE OF CONTENTS

"Western Knowledge" is a term that inspires confidence in most people. University scholars have decided on the slow evolutionary history of man. Via social propagation, this permeates all aspects of society. Agencies like NASA carry a message implying that we know about our planetary system and the history of the human race. To offer any alternative calendar or scientific belief is not allowed. But there are major contradictions that cannot be ignored. As we look at history through the eyes of the ancients, they talk figuratively about giant comets and major earth changes with the death and destruction they wrought. They portray the past through Myths Magic and Mayhem. Our history as taught could not possibly be true. This begs the question …

What is … The Real History of Man & Earth.

INTRODUCTION

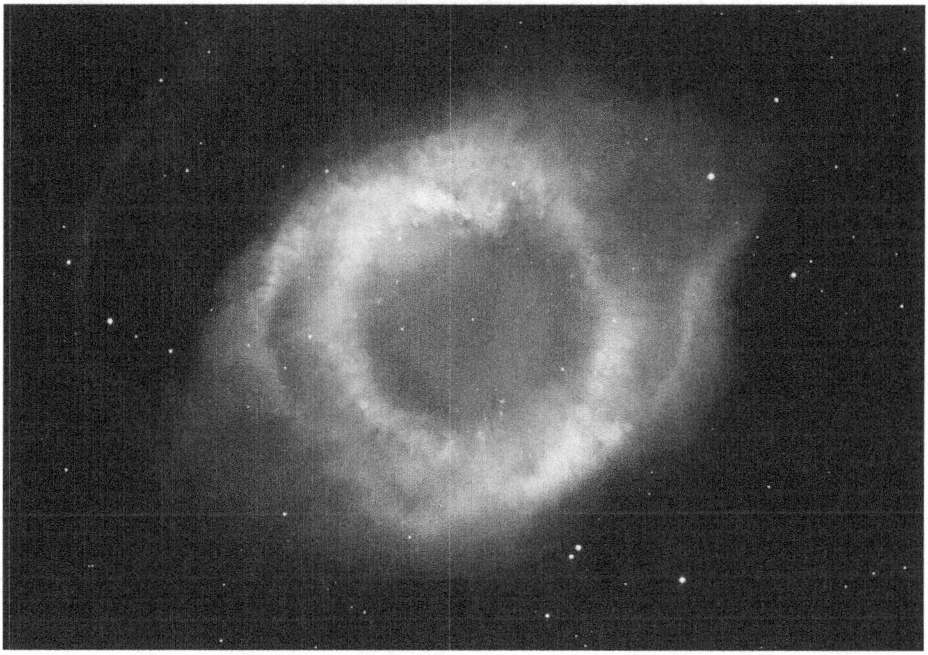

The above photo is of one of millions of stellar "nebular ion clouds" surrounding stars that cannot be explained by standard astronomical theory. If you could travel a short distance out of our solar system and look back, you would see our sun to be very similar. It is a cloud of dust and gases held back by a positively charged proton wind that emanates from the fusion based star (what we call the "solar wind"). This creates an electrical capacitor that is instrumental in causing a myriad of effects and additionally provides the energy for comets to become planets (as explained in my Plasma Discharge Comet Model) in a complex cosmic game of celestial billiards. This book discusses how these physical processes have drastically affected the History of Man and Earth leaving us a view through the eyes of the ancients of Myths Magic and Mayhem.

To put this photo into perspective, the center dot would be our sun and planets out to about Earth. The inner part of the nebular ion cloud would be about where the planet Pluto orbits the sun. The region with brightly colored nebula would be what astronomers call "The Kuiper Belt" where tens of thousands of forming planets have been discovered in recent years. You can see the large comets forming planets. One by one, Jupiter and our Sun (a twin star) captured all of the planets and moons into the solar system by a gravitational process known as "capture." All of the existing planets initiated in the form of large planetary sized comets. My science describes the physics of these processes in my "Plasma Discharge Comet Model."

This model shows large comets growing into planets as the tail material is pulled ***into*** the comet nucleus by powerful electrical forces. It is this same influx of matter that causes what I term "tail drag." This causes the newly formed planets elliptical orbits to become circularized and fall into the orbits around the sun that we see today. During the planet capture and formation processes, the huge comets can cause great devastation to the existing planets like Earth (sometimes during multiple passages). They do not collide with the existing planets, but pass nearby. This scenario explains the many earth effects not explained by a single small comet impact whose energy is absorbed locally.

"Action at a Distance" is the term I coined for the

thousands of effects that can take place when a large planet sized comet nucleus and its electromagnetic plasma sheath passes near Earth or other planets like Mars. The ancients tell us through their writings and petroglyphs about their experiences with such events. Man's development and Earth itself have been greatly altered by these events. My prior books go into great detail about the Earth Changes and effects that will only be briefly summarized in this book.

They include the following: Pole shifts, mountain building, rapid plate movements, massive earthquakes, worldwide volcanic eruptions, tsunamis in both the oceans and land, blocking sunlight causing rapid radical cooling, glaciation, ice ages, severe weather, influx of materials from the comet tail including water, hydrocarbons (oil what the ancients called naphtha), meteors, dust with high concentrations of minerals, nano-diamonds (which are now known to form to the sunward side of comets), worldwide auroras and upper atmospheric electrical effects, huge vibrational sounds (the ancients called trumpeting), electromagnetic effects including comet to planet discharges, and the list goes on. Large influxes of space dust can cover existing man made structures with dust and sand in a single day long passage. This makes more sense for sites such as Gobekli Tepe and hundreds more than to believe that the inhabitants buried their own thriving cities.

This leaves the earth reeling for decades and centuries to come. The ancients who survived such events pass down their experiences attributing them by way of Myths Magic and Mayhem. A single small comet strike cannot explain these same effects which are almost universally observed in the geological record. The average groups studying human history today, which includes professional astronomers, archaeologists and independent researchers, are so affixed to the "an asteroid or comet hit it" concept that they have no idea that it does not explain the myriad of effects we observe in the hundreds of locations. Additionally, the earth has experienced many hundreds of such events, not just one or two.

My book "**Principia Meteorologia - The Physics of Sun Earth Weather**" (2004) details the effects of the passage of a large comet not only on planet Earth but other planets such as Mars. Mars shows that it recently had an atmosphere and oceans that were removed as a large comet passed by and removed these as observed by the ancients. The original book of this library series, "**Planet X Comets and Earth Changes**" (2000) as well as the follow on books CDs DVDs and Lecture Series "**Surviving Planet X Passage**," "**Atlantis to Tesla – The Kolbrin Connection**," "**COMETS**," my 18 Hour MP3 Lecture Series "**The Physics of Ancient Celestial Disasters**" and the 180 Hour Lecture Series "**Science Home Education Package**" further detail the history of man and Earth. The current book is a summary of the various aspects of these works and is being released at a time when the public is finally waking up to the reality of multiple recent massive earth wide disasters and dispelling the slow evolution history promulgated by NASA and standard astronomy.

This brings us to some major conclusions. Man's history has had MANY Earth-shattering events including near mass extinctions. This book describes these events as related to the physics of comets and solar system formation. Earth has been assaulted and re-layered many times in its history. That is why the more we dig, the more we find unknown ancient ruins of lost civilizations scorched and buried in the aftermath. The true age of planet Earth and the other planets of the solar system cannot be known because the surfaces have been drastically altered many times. Dating material found on the surface, as has been done, will not produce correct date results.

Researchers who still believe that extinction event Earth Changes were caused by small celestial objects colliding with Earth, must see that small asteroids or comets colliding with Earth do not explain the worldwide effects we see. For example, using my Plasma Discharge Comet Model and laboratory experiments, micro diamonds form to the sunward side of the plasma discharge comets just as diamonds form in the laboratory vacuum chamber with carbon rich gases in an electron current. During a close passage of Earth by a large comet, uncountable numbers of micro diamonds rain in along with other forms of cosmic pollution.

The idea being proposed today by researchers (both from the archaeology and science astronomy worlds) that a small comet hitting Earth would create millions of micro diamonds is simply incorrect. This book addresses these issues in detail. Two points are clear. First, devastation at the hands of a large passing comet has happened many times in Man and Earth's history and secondly, buried under repeated layers (not just one or two) are the remnants of many deceased civilizations of man. The history of Man on planet Earth may go back millions of years or even much more. The Real History of Man and Earth are far different than what you have been taught. This history is filled with Myths Magic and Mayhem.

The main reason for releasing this book now is because there is so much controlled misinformation emerging now regarding two events called the "Younger Dryas Event" which researchers claim occurred some 12,800 and 11,500 years ago. They claim that a small comet hit Earth and caused all kinds of unrelated events. One great problem is that dating techniques are more of a bandwagon consensus rather than based on sound science. Unfortunately, many researchers are jumping on this bandwagon. Here are some main points. First of all, this has been thoroughly presented before many times starting in 1979 by two astronomers Clube and Napier who envisioned multiple impacts..

They sold a lot of books and gained fame only to have this fall by the wayside. Secondly, and most importantly, a single comet hitting Earth is assumed because it is politically correct (it agrees with the NASA propaganda that comets would be harmless little snowballs) and thirdly, none of the worldwide effects they are claiming could be the result of a small comet impact. A side issue is that we know there were numerous devastating events much more recently than the out of sight out of mind 12,000 year dating scheme. And lastly, they cannot find the crater associated with such an event that would certainly be very visible on Earth's surface. Every 10 years this same scenario rises out of the ashes. The problem is that the real reasons for what we observe in the archaeological record have been established as being caused by large comets passing by Earth by "Action at a Distance" effects that do in fact explain what the ancients describe (the huge comets in the sky that attack Earth and pass by some to return again another day).

Decades ago, there was a group claiming that the dinosaurs went extinct in a single comet strike of Earth. They claimed that the crater was found in the northern region of the Yucatan peninsula in the Gulf of Mexico. The fanfare was heard worldwide and is unfortunately still repeated as some kind of established fact. This was finally laid to rest when diligent researchers studied the animal life below and on top of this crater (which was in fact caused by such an asteroid or comet impact). They found that the dinosaurs did not even die off locally after the impact let alone worldwide.

Why this debunked scenario is gaining air again speaks of a giant disinformation campaign. We are finding out that some people are receiving money under the table to "continue their research." It is great for their book sales, speaking tours and the fame and all that goes with it. Below the surface however, this speaks of a well funded misinformation program to popularize a completely improbable scenario while squelching the real story from the public. You are already seeing

Netflix and History Channel specials and blanket coverage in the international fake news media. This is clear evidence of a massive well coordinated distortion that is designed to divert the public from real science, real archaeology and the true history of Man and Earth. Today man is at a critical cross roads and researchers who present mountains of good information but come to completely false conclusions (while greatly profiting) are useful pawns in this deadly game of science and history distortion.

These stories gain momentum and it becomes fashionable to jump on board. There may even be a Nobel Prize in it for some of the more aggressive Ivory Tower supporters of this movement. Objections are drowned out in the rush to fame and blanket coverage in the controlled news media. Those partaking in this charade love the attention while always believing they have hit the mark and are changing the narrative for the advancement of the human race. The easiest way to control most people is via egos.

I am releasing this book now in an attempt to reign in on this disinformation campaign (just as I have done in the past with the bogus CIA Planet X intel program, the Millennium Bug scare, the 2012 Mayan Calendar baloney and countless other CIA misinformation programs).

Since I was a young person living in Mexico City (early 1970s), studying both archaeology and the physics of solar system formation, I have been involved in these issues. I have been involved in this long before most of the current crop of "researchers" were born.

I have seen misinformation come and go, and also witnessed the distortion of history as academicians, archaeologists, physicists and astronomers, geologists and many other professionals amassed volumes of information but used "accepted science" to come to completely incorrect conclusions.

To gain pubic view in the controlled media, one has to be politically correct relative to standard astronomy claims of how the solar system works and its time frame based on slow cosmic and biological Darwinian Evolution.

The biggest mistake made by all is to put dates on events. Dating is nearly impossible, yet the very first thing most researchers do is to put a date on a place or event. They argue against current dating structures, but immediately enter their own dates. This only serves to confound the real history of man. This book hopefully will put some sanity into the current fray of misinformation hitting the public in the face today.

Today independent researchers are breaking away from the archaic ideas of archaeology stating what should have been obvious from the very beginning. Many early archaeologists had the concepts correct (that there were other far more ancient and advanced civilizations on earth that created the amazing stone work). The interpretation of ancient architecture and building techniques are not explained with standard explanations and one has to make the connection to a much deeper past and earlier technologically advanced civilizations. These same researchers unfortunately then cling to archaic outdated astronomical concepts in attempts to explain the celestial influences that are visible at the countless ancient archaeological sites. It is time to also step away from outdated astronomical theories and recognize that our universities, museums and planetariums are filled with "textbook repeaters" still propagating ideas that could not possibly be true.

I have a suggestion to all researchers and interested parties. First, let us remove the dating structures both old and new because literally we are just beginning to unravel the true story. Secondly, let all involved in this renaissance of archaeological discovery come to the conclusion that standard astronomical theories (as promulgated by Carl Sagan in his over rated COSMOS series) are completely erroneous and a completely new set of astronomical concepts must take their place based on my extensive reconstruction of the workings of the cosmos and the Plasma Discharge Comet Model.

Below is another example of what I call in my theoretical work the "Nebular Ion Cloud." This is an angle view of the plane of that planetary system. Here too you see the large comets forming planets (with their Plasma Tails pointing away from the central star). This region is the equivalent of our Kuiper Belt where thousands of new planets are forming.

When the New Horizons space craft arrived at the planet Pluto, which is on the inner edge of our Nebular Ion Cloud, it saw the comet tails of both Pluto and its companion Charon. They also show photos of the comet tail material falling into the planet as ammonia and other nitrogen compound "snow" but as with all data NASA could not explain it so they ignored it. My book library and lecture series, written for the general public and professionals alike, are available and must be read to put this book into perspective. They detail my theoretical physics developed 1979 to 1981 as I was on the faculty of both the Physics and Mathematics Departments of Cornell University.

This replaces the NASA dirty snowball comet model and other concepts of astronomy that have been thoroughly dis-proven by space craft data. My books and lectures library make a wonderful gift for youth, their parents or for serious researchers who want to know the real story of the cosmos that we live in. Follow the link below for more details.

https://www.jmccanneyscience.com/

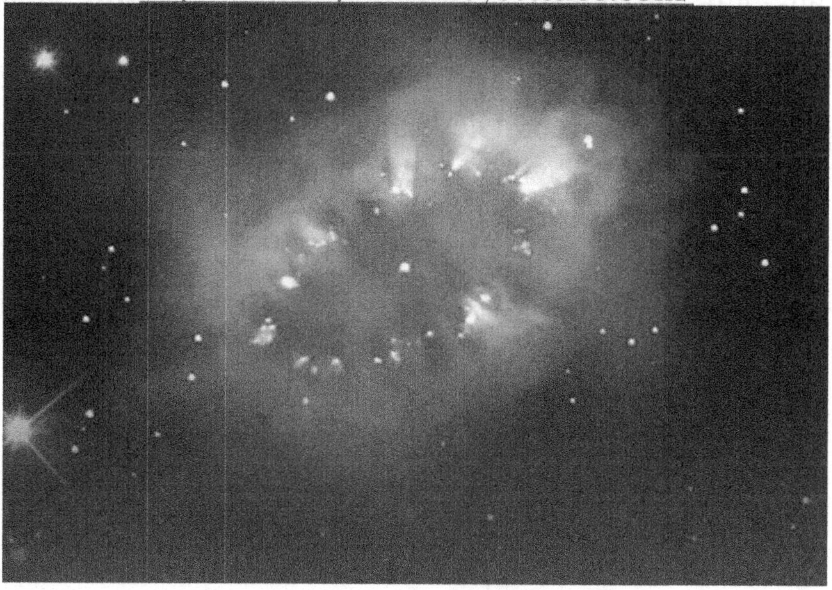

FOREWORD

The Real History of Man & Earth Myths Magic and Mayhem

This book is based on past works that must be read to put this book into perspective. When Solon, the great uncle of Aristotle, went to Egypt as a young scholar seeking knowledge of ancient history from the Priest Cults, they stated that the Greeks descended from sheep herders that survived an amazing Earth-wide disaster. They also recounted that there was an amazing advanced civilization called Atlantis that went under the ocean in a single catastrophic event and that the prior Greek armies were at war with Atlantis when this occurred. They told Solon that no less than five major Earth wide catastrophes befell them in the prior 10,000 years. Every ancient civilization has similar stories.

In the 1800s as the white European settlers populated the central and western states of the USA, they all recounted amazing destruction resulting from a great flood. By the time the 20th century came with the industrial revolution, all remnants of the flood had been eliminated by man. Ignatius Donnelly, the Lieutenant Governor of Minnesota wrote a book "Atlantis: the Antediluvian World" (1882) on the knowledge of then discovered artifacts that pointed towards the extensive great flood. He describes how a farmer came to him with a human doll found while digging a new well. But modern "scholars" call his work pseudo archaeology.

In 2003 I published a book of my studies dealing with Atlantis and the man named Nikola Tesla … "**Atlantis to Tesla - The Kolbrin Connection**." This summarized the discovery of the original site of Atlantis and exposed that it had been under excavation by the US Navy in its resting place of the South China Sea. But Atlantis was just one chapter in the many that unfolded as we opened the pages of those left by the Maya, Olmecs, Incas, China, India, Egypt, Natives of the North American continent and more. ALL had the same story. Mountain building, great floods, tsunamis caused by passing celestial objects. Another of my discoveries related that the Nazca Lines figures were an ancient antenna complex capable of receiving and sending signals to distant star systems.

Many modern writers had their lives destroyed because they put forth information that contradicted the status quo. I, myself, felt the wrath of their heavy hand of righteousness as my career was cut short and work suppressed. In 1979, while I was a faculty member of the Physics and Mathematics Departments of Cornell University (Ithaca, NY), a man named Carl Sagan (then David Duncan, Chair of Astronomy also at Cornell) spearheaded a television series called COSMOS. Co-written with his wife Ann Druyan, the series was to set the direction that all science would have to follow for all future time. Sagan talked about open discussion and search for knowledge, but there was no one more responsible for the lock on alternative thinking than Sagan. The science news media has been completely sequestered into promoting "standard western knowledge" and making sure that no one else can ask a question or breach a topic opposed to what has really become a religion. All subjects have been synced into a lock step with Sagan's outline including archaeology, anthropology, geology and planetary science.

"Western Knowledge" is a term that inspires confidence in most people. University scholars have decided on the evolutionary history of man. Via social propagation, this permeates all aspects of society. NASA carries a message implying that we know about our solar system and the history of the human race. To offer any alternative calendar or scientific belief is not allowed. But there are major contradictions that cannot be ignored. Our history, as taught, could not possibly be true. This begs the question … What is … The Real History of Man & Earth

Α Β Χ Δ Ε Φ Γ Η Ι ϑ Κ Λ Μ Ν Ο Π Θ Ρ Σ Τ Υ Ω Ξ Ψ Ζ

PREFACE

"In the beginning there was light" … is how it all began … then in the end He finally created MAN in his image and likeness. The Book of Genesis. There is evidence that man has been here on Planet Earth for 200,000 years and possibly a lot longer. Did any of man's successful civilizations have space faring advanced technologies that went to an untimely demise? The human race collectively has been very resilient. But is it really? Or are we living on a razor thin edge between the future and mass extinction?

Other species like the dinosaurs, woolly mammoths and mastodons, and dozens of other examples of smaller creatures on Earth have met with mass extinctions. Huge Earth changes have occurred at the same time. The face of Earth was altered seemingly overnight. Mountains formed with sea shells and sedimentary rock pushed to the highest peaks, while existing mountain peaks were thrust under the deepest parts of the ocean. Atmospheres were completely changed and poles were shifted to give new orbital parameters to Earth's path around the sun. NASA and standard science brush this aside, always claiming that an asteroid hit Earth while sweeping the real data under the rug. The reality is that many major catastrophes have happened as large comets came nearby Earth, causing what I call "Action at a Distance." A single asteroid collision does not explain what we see when we look at all of the data.

There is a chapter included in all of my books dealing with Astronomy, Astro Physics and Space Science entitled "Will Man Survive?" that is a parody or puzzle for which everyone can make his or her own conclusion about the survivability of man. The fact that intelligence agencies have completely blocked real information from the public about one of the most important factors in our very survival, tells you that the public is not very scientifically astute and that there is a fear of the public becoming aware of the truth. The same controllers continually fabricate stories that completely captivate and control the public.

When we can go into space and envision travel to distant stars, is it possible that man has done this before or even arrived from another distant star system? Or have we been a captive species just recently emerging from the cave ages with an ape as our only ancestor? Is it possible that man did not evolve from any other "species" on Earth but has been in his current form for millions of years? Is it possible that Darwin's evolution only makes sense if one assumes that all life slowly evolved on this planet? Is the reason we cannot find the evolutionary "missing link" because Darwin's evolution is flawed? Could it be there is no missing link? The reality is that no one is allowed to think about such things in today's government funded ivory tower bastions of "western knowledge."

I discuss this very topic in my book "**The Diamond Principle**" (2010) and discuss the additional element that the human race, in its current condition, is not psychologically prepared to go into space. We lack the collective understanding of all of the thousands of factors that allow us our comfortable existence on this lovely blue water planet. Most people think of space travel as a technological feat. Western society collectively lacks almost all of the skills that it would take in the spiritual and mental sense. There are many open questions related to man and his history on planet Earth. They all involve things from outer space. Once one starts to peel away the thin first onion layer of standard "western knowledge," it comes apart like a newspaper on a rainy day.

All of these questions are going to be raised and more in this long-awaited text. The only conclusion we can reach at this point is that what is currently being taught as the history of Man and Earth could not possibly be true. It is OK to say "we don't know" and "we do not have all the answers" rather than rush to conclusions that take more time to undo than they are worth.

I.
PLANET X, COMETS AND EARTH CHANGES

News Flash! Comets are not Dirty Snowballs!

Introduction to: The Plasma Discharge Comet Model and The Electrical Universe principles as related to Comet Hale Bopp and Planet X – Hale Bopp is the newest member of our growing family of planets

If there were any phrase that must ring out loud and clear and dominate the pages of this book it is the above statement that "comets are not Dirty Snowballs."

Space probes have visited a dozen small comets including landing on the surface of one comet nucleus. The evidence is conclusive ... there is no snow or ice or water in any form. What they found was a hot dry surface of organic sand and in some cases tar like black substance (hydrocarbons) baked onto the hot dry surfaces. Additionally, there is no water coming off any of the comet nuclei nor from any subsurface water sources.

Early NASA artists' conceptions showed an anticipated snowy surface with massive geyser-like water sprays jetting into space; however, reality shows none of this. The European Space Agency lead scientist publicly stated that they disproved the "dirty snowball" model, yet the fake science news media continues to push this false narrative. My other books discuss this topic in detail. Suffice it to say for the current text that NASA truly means Never A Straight Answer. Comets are electrical discharges of the solar capacitor as proven in my Plasma Discharge Comet Model. We also now know that comet nuclei can on rare occasions be planet sized.

It would seem that an innovative theory that actually explains space probe data and which actually explains not only the physical past of our solar system and planet, but explains man's historical interactions with formation of new planets, would be welcomed and hosted as possibly one of the highest achievements of man's quest for knowledge.

Just the opposite is true. Over twenty years ago at the time of this writing, the "**Planet X Comets and Earth Changes**" book made public my work on what had been previously only well known in higher level scientific circles. The book would have made the New York Times best seller list but since it was only distributed through private channels it saw no outwards fame.

If you have reached this point in this book without reading the Planet X and other related books, then you may proceed but with the caution that you cannot make any real informed decisions without reading the prior books. Remember that these are only small public releases compared to the much larger quantities of material that exist in a more scientific presentation. The form of the current book is in written introductions for each chapter, which then links to recorded lectures. It is certain that if "professional" astronomers get to this point the usual response is that since this does not fit into their assumed paradigm then they can rest on their more than lazy laurels that have brought us to this point in the first place.

This is a topic so complicated that written text cannot serve to explain all of the interwoven details. "**Planet X Comets and Earth Changes**" set the stage for a lot more than just presenting a "comet theory." It relayed decades of past publications by this author and also presented a complete theoretical picture of the workings of the solar system and even the galaxy in which we reside. The timing for its release threw a monkey wrench into the planned disinformation plan of the CIA controlled NASA. They came out kicking and screaming. Things were about to get very ugly in a history of a topic that already had a long history of ugliness.

It was timed with the middle of one of the most egregious disinformation campaigns ever hurled onto the mental scene in the history of mankind. The Planet X insanity raged as a crazy lady who claimed she talked to aliens about an impending worldwide disaster to be caused by the infamous "Planet X" on May 15, 2003. NASA supported her behind the scenes from the hideous inception of the disinformation (helping to select that date that turned out to be an unusual orange full moon) to the constant support providing photos of distant star fields with the alleged Planet X (allegedly taken by the crazy lady's companions the alien "Zetas," which she claimed she "channeled with" from the relatively nearby visible twin star system Zeta Reticuli).

What is it that they were hiding? In the face of clear evidence that comets are not small dirty snowballs, why was the controlled fake news media continuing to push the false narrative? The current book discusses these issues in detail.

II.
SURVIVING PLANET X PASSAGE

The second book of this series (a 60 page pamphlet) "**Surviving Planet X Passage**," speaks directly of the Earth-Changes events that we know have happened in the past and that certainly someday will happen again. What is interesting is that in the original text I explained how to build a survival calendar from a pile of rocks. Not only was this important for the sake of survival but after this I started receiving notices of archaeological finds from all over the world where such piles of rocks have been found. So the ancients knew of the skill and indeed many used it to determine the planting seasons and for other reasons, as I state in the pamphlet.

I have since produced a product called the Emergency Survival Re-Calendar Sun Star Clock, which is for sale on my web page. The idea was that we very much lack something in our study of artifacts from the ancient past as we dig them up from every corner of the globe. We lack the information on the location and orbits of the planets during past epochs.

This is important for the same reasons that Immanuel Velikovsky was studying ancient calendars. It provides us with not only a description of how things were before but also provides dates when things changed. So, without such information the astronomical community can judgmentally proclaim that the solar system was always as it is now ... without a shred of proof ... and by the same token no one can prove them incorrect. It is like so many other "theories" of astronomy, such as the famous "Oort Cloud" where allegedly comets reside. The only problem is that no one can ever detect it and therefore no one can make measurements to disprove it. Scientific inquiry should rightly prohibit such claims but they do not.

Another example of such unfounded claims is the Darwinian theory of evolution. Today scientists are still looking for even a single "missing link" and cannot find even one ... whereas there should be tens of thousands of them looking at us out of the dust at every corner and turn. With absolutely no evidence, the theory of evolution binds the sciences of organic and inorganic chemistry together. Scientists cannot explain why a carbon atom in an inorganic molecule acts differently than the same atom in an organic molecule, so they simply state that one is organic while the other molecule is inorganic. They cannot tell you any more than that.

The survivability of man is one of the most essential aspects of our existence, but the nightly news is filled with topics that are on less than a few days' worth of impact. This is an essential book to understand the level of severity of the largest natural disasters that have and will befall mankind.

III.
ATLANTIS TO TESLA - THE KOLBRIN CONNECTION

This third book in the series rode the momentum wave of the prior two books but with its presale announcement everyone was expecting another Planet X related book. When the title was read on Coast to Coast AM, the first three printings sold out in about an hour. There was about $42,000 in sales by the time the sun rose in the morning. It created a problem with my internet merchant account because they suspected some kind of fraud since it outstripped any of my past sales by a huge amount. They required that I fulfill all the orders and verify that the customers got the books before they would release the funds. This took a few weeks and thus ended that uncomfortable issue. My merchant account raised my limit to $50,000 to accommodate any future sales spikes.

This book contained hot buttons for every late night radio buff. It would be like a new James Bond thriller but in the world of late night talk show mindset. Tesla – Atlantis – Secret Societies – ancient technologies –space travel via electromagnetic propulsion systems used by the Atlanteans and, newly discovered as I approached the military industrial complex — the newly released KOLBRIN by the Culdian Trust, and the list goes on. One person I talked to nearly daily, who listened to most talk shows and informed me of their various discussions, told me that I had created more careers with this book than any other topic.

Pretty soon everyone was an expert on all of these topics, not the least of which was "secret societies." These topics dominated radio and internet as well as publications of all varieties. Something happened also as this book hit the public. There was a group of military experimentalists who setup a Tesla Tower to attempt to "drill into the ionosphere" as I had described. They succeeded as the tower was constructed near a nuclear power plant outside of Ottawa, Canada. They were going to use the nuclear plant to power the tower to extract energy from the ionosphere, as I had described. They did not have enough power from the local plant so they requested more power be switched across grid elements from the USA across the border. When the current surge came down the path from the ionosphere to the station, the surge blew through the protective circuitry and backed up nine nuclear power plants in the USA northeast sector.

This was the famous northeast power outage that destroyed electronics in everything, including automobiles. The reason given was that a squirrel's tail fell across a power line in Ohio and caused the major outage. They did not try this little experiment again.

This book connected the physics of planetary formation with the useful aspects of electricity that comes from outer space and lead to other books such as the "**Principia Meteorologia – The Physics of Sun Earth Weather,**" which outlined my lifelong work on weather and the electrical nature of our solar system.

There is a series of other books that extend the topics raised in these first three books. These include "**The Diamond Principle**" and the series of texts dealing with prime numbers and computer encryption, illuminating the secret and deranged world of cybersecurity. There are many other positive scientific efforts underway. See the web page for details and more information www.jmccanneyscience.com

IV.
ABOUT THE COVER PHOTOS

The image is of a human face with the mind reaching to the stars. There is a 1960s song "The Age of Aquarius" that states we are heading into an age of enlightenment and associating us with the stars. Imagine living in a star cluster farther in towards the center of our galaxy where all you could see would be the hundreds of local stars shining brightly in the sky. There would be no night sky at all since it would be filled with bright stars no matter how you were positioned.

On the other hand, the benefit would be that we would probably become space faring and visit other stars much more readily than we would. Residents there may never know that there is an entire galaxy and universe with billions of galaxies beyond their own. But no matter where one lives in this big galaxy, the fascination of looking up into the sky would be the same. At this point we have to come to a single realization. That realization is that "MAN," as we are on planet Earth, is not just "an intelligent being" with a bigger brain. MAN has a spirit that is living in each and every body and is well connected to the universe. This consciousness is essential to the makeup of man. This is what makes man different from the animals that only react to their environment. The cover photo depicts the spirit of man.

In the forehead is embedded a photo of comet Halley taken by the Hubble Space telescope during this comet's passage in 1986. Halley's comet is a very small comet and certainly nothing that would be on our list of objects that could grow to become a planet. But as you will discover in this detailed analysis of the history of man, planetary sized comets play a decisive role. Man's spirit knows and fears comets like a fawn deer fears the smell of man.

The reality is that comets are usually small and because they lack gravitational field sufficient to hold matter in their vicinity, they usually just continue to be jostled by electric currents. Some material is baked on their hot tarry surfaces while other surface materials are eroded away with electrical discharges burning them during close passages to the sun.

Large comets can become planets under very specific and rare conditions. It is amazing that man instinctively harbors a fear of comets. The one thing that is common to all ancient legends is the notion of electrical discharges being hurled amongst the planets.

The idea that comets would have a discharge was accentuated by the 1950s work of Immanuel Velikovsky in his discussions of the Comet Venus becoming the Planet Venus and that literally every ancient culture made more than passing reference to this event. The astronomical communities had no more than made their initial defenses of such "nonsense" when comet Arend-Roland came through the solar system in 1956, sporting a brilliant sunward spike that only should have bolstered the idea of electrical discharges in comets. But, instead, the astronomical community claimed that this was just an optical illusion. Nothing to see here; please move along.

When we talk about large comets and their effects on existing planets like Earth or Mars, at the heart of all discussions has to be the subject of electricity from outer space. My Plasma Discharge Comet Model is based on the fact that electrical conditions are dominated by the Solar Wind's creation of electric fields that span our solar system. In 1996 the first official measurement of a comet exhibiting X-Rays was recorded. Comet Hale Bopp was the second comet to be viewed with x-ray telescopes and this again was the last thing the CIA controlled NASA scientists needed to explain with ice ball comet theory. One of my main predictions in my 1979 published papers was that comets would produce x-rays just as discovered.

It is beyond bizarre that today we have data from a scientific payload that landed on the nucleus of a comet, which shows that there is no ice or snow or water in any form on the comet nucleus, yet the "fluffy little wanderer" dialogue continues in the fake science press. Now snowballs produce x-rays.

There is wall to wall convergence of new and old comet descriptions in the tier 3 scientific press written for the public with never ending stream of phrases referring to terms that makes one think that this is somehow still true. How can the public be so misinformed and continue to be blinded in ignorance?

Only a scientifically inept populous could continue to be fooled in the face of overwhelming evidence to the contrary. But what about scientists themselves? All I can imagine is that the thought of no retirement or loss of tenure or publication access is enough to make them all mum as church mice singing to their respective choirs.

Before the public can understand the issues related to the history of man, they must understand the major forces at play in the solar system and universe in which we live. The reason I mention the CIA and NASA is because they have been essential to the catering of false information to the public about the real issue of space. Keeping the general public ignorant has been at work behind the scenes by those in control for a long time.

If the public realized the dangers that lurked in and around us (not just little colliding asteroids) in the form of large comets, they would instantly understand that the "leaders" had no ability to do anything about this and the "leaders" would instantly lose their power over the populous. Although my work has been central to many controversies over the past half century, it took me a long time to realize what the core issue was and who was actually pulling the strings to keeping my work suppressed.

Here is an experiment for anyone doubting this suppression. Go to any YouTube channel (pick one dealing with weather or space or geology or archaeology, etc.) and type in the blog notes below the video a favorable comment about me or my web page. You will find the following. When you return to that YouTube video with the computer you used to make the blog note, you will continue to see it there. Now go to another computer and log into the same YouTube video. Your posting will not be there.

You then have to ask yourself ... with all the insanity and bad science and nut cases all over the YouTube platform, including people copying my work, why is my name being shadow banned? You are now seeing that the information I am pointing out is correct and in direct conflict with the misinformation campaigns that are being pushed onto the public.

I also include a photo page with a nebular ion cloud and explained how you can see planets forming from comets around other stars. The same process is working in our own star system. This is essential to understanding past devastating events on planet Earth and our nearest neighbor Mars.

This brings up the final point that if the human race is to survive, we have to learn to be good stewards of this planet and also to move large populations into outer space. If man does not move towards major movement of population into space, he is doomed to mass extinction. It may not happen in your life times or even your children's lifetimes, but it will eventually happen. When it happens, it may not create a complete mass extinction, but maybe it will.

V.
STANDARD SCIENCE and the PYRAMID OF KNOWLEDGE

That old saying “he who laughs last laughs best” may be applicable to the world of standard science and the overseeing powers that control the myopic residents of our major universities and seats of knowledge. To understand the conundrum that we face as independent thinkers today, one only has to look at the history of man to see its many empires that have come and gone. In the heyday of all of these, there were people who were capable and built them to unlimited heights. After a time, the people got fat and lazy living off the fat of the land. The follow on leaders became slothful and delved into the seven capital sins as they lived lavish life styles.

Today is no exception. Today’s academicians’ life … what I call the “scientific vacation” … abounds with benefits. Like pampered movie stars they fly around the world on someone else’s credit card defending their realms of knowledge. This would all be well and good if they were not carrying belief systems that were conceived long before the space age began. Literally all scientific theories were created decades ago and set into stone. I call them the “Axioms of Astronomy.”

They include but are not limited to the following. If you read my work and listen to my extensive lecture series you will learn that none of these could possibly be true. They are noted in the following bullet points:

- Comets they say are dirty snowballs … little fluffy balls of snow and ice and dirt that lose mass when they approach the sun. (They also claim that these formed in a cloud called the Oort cloud that has never been and never could be a detected … but almost all of their theories are based on baseless assumptions like this.)
- They claim that planetary systems formed all at one time and aside from a rare asteroid collision, nothing has happened since to the individual planets.
- They say that space is electrically neutral and believe that plasma magnetic fields just exist on their own (a contradiction of basic electromagnetic theory).
- They claim that the universe began in a big bang about thirteen billion years ago and everything we see is just the result of randomness (there is no grand design).

Sub corollaries of this archaic belief system are the following:

- Their beliefs bring them to conclude that everything you see happening on planet Earth has to be explained by "evolution" and natural causes only on planet Earth … there is no room for any outside influence except for a few minor exchanges with the outside environment.
- This means that the mountains, oceans, landscape in general, atmosphere, life,

EVERYTHING is the result of Earth doing all of this on its own.

- Their belief system then takes you to other members of the solar system and in fact the galaxy and they claim that these too are all isolated … all having been formed billions of years ago and all evolving completely separately.
- The biggest part of their belief system is that all of this just happened by itself and there is no underlying design.

Other theories on how things work include the following:

- Plate tectonics they say causes the plates of the Earth to move around sideways by simply sliding using something called subduction. There is no explanation in terms of physics of how this amazing feat is accomplished or how it somehow magically creates mountain ranges thousands of miles away.
- They claim that comets brought the oceans of water to Earth. But now we discover, as I predicted, that comets have no water on their nucleus and, even if they did, a direct collision would expend so much energy that it would vaporize any water sending it into outer space and would not provide Earth with its supply of water.
- The same with the atmosphere there is no explanation for the source of our atmosphere or why it has changed multiple times (at the same times as mass extinctions on Earth).

I could go on and on but all of this and more is covered in my published and readily available work and lecture series. The point is that this "pyramid of knowledge" has astronomy at the pinnacle with all other subjects falling in line below it. None of the other subjects can alter from this cast-in-stone set of rules. These basics hold no water at all; however, they are maintained by tens of thousands of what I call "Tier 2" scientists who convene semi-annually in meetings like the American Geophysical Union and hundreds of other private meetings. Advancements are painstakingly slow as the so called "respectable peer review journals" pages are filled with never ending Babylon of incomprehensible language.

The biggest issue is that all of this is 100% funded by federal funding and NO ONE—but NO ONE—is going to rock the apple cart to disagree or throw stones at the roots of this more than beneficial system for its members. To disagree would mean immediate disdain from ones peers with loss of funding, publication access, and his or her career. Where would any of these people get a real job? Their lives would be ruined. It is a closed system that cannot and will not change. It is impermeable to outsiders who are scoffed at.

One of the main living mouthpieces for this establishment science is Neal de Grasse Tyson who is curator of a prestigious planetarium, who marches the party line and keeps it in order. He recently presented an "expert class" on an internet platform and said the following in selling his product. (Paraphrased) There is such a body of knowledge that supports our position (of which we have a monopoly and no one else is allowed into our playpen) … and because of this … no one else will ever have this amount of information and knowledge … so therefore we can never be challenged and any challengers could never be correct in their challenge." Yes, it is a fully closed loop system.

No one else can use the equipment. No one else can see the detailed results. No one else can propose theories and no one else can question the self-ordained correctness of their theories. All of this is happening as they taut the open dialogue of "science" and the "scientific method." It is all a giant sham.

This is the pyramid of knowledge that controls everything you will ever see in the western press or other news outlets. The only other issue (to be covered in a future chapter) is the controlled opposition. Behind all that we see is a well orchestrated misinformation program in which both sides of any question are controlled. This is under the control of the CIA (police force of the world bank) who took over NASA and fully controls the information that reaches the public (see my book "**Planet X Update 2020 – The Real History of Planet X"**).

Along with this is the support of controlled opposition. These are the nut case impostors that are supported under the table to create goofy "alternative science" so that the white hats like Neal de Grasse Tyson can ride in and save the day. Real information from real scientists, therefore, never sees the light of day.

VI.
AN ASTEROID HIT IT?

No sooner was the ink dry on the paper where NASA and the Carl Sagan entourage of scientists buried Immanuel Velikovsky at the 1974 AAAS meeting (called for the express purpose of doing just that), that a Nobel prize winning physicist proclaimed that an entire worldwide mass extinction occurred because of a collision by a comet with Earth. Sagan and the boys should have had egg all over their collective faces but it was at this juncture that they decided that if anything unusual happened that could not be explained by slow gradual processes, they would allow that "an asteroid hit it" could be used to cover any such defugalties. You would not believe all the times this has been implemented since.

Literally every time an event occurs in space (e.g. a comet brightens unexpectedly) the news media simply adds "an asteroid hit it" and who would be any the wiser? It is something that can neither be proven, NOR DISPROVEN. It has been used to explain unexplainable magnetic fields, explosions on Earth that have no remnant of asteroids, such as the Tunguska event and, as I said, thousands of other incidents.

More recently nonstandard scientists, as with Clube and Napier and thousands of other independent "scholars," have fallen back on the "an asteroid hit it" explanation for the demise of species to causes of ice ages and Earth climate changes. The issue has always been with any of these "an asteroid hit it" explanations that even very small localized asteroid collisions leave an indelible mark on the surface of the Earth. One of the most famous is in Winslow, Arizona. In the middle of nowhere there is a relatively small crater. The remnant that hit there is on display in the associated museum and sets on top of a pedestal. Although the crater is more than a kilometer in diameter and 175 meters deep, the attacking asteroid is surprisingly small (but it is there and not that difficult to find).

You can see that the debris field surrounds the spot and all of that land mass had to be blown out over the surrounding desert. (You would not have wanted to be in the area when this hit). But in terms of Earth wide reaction, the bulk of the impact WAS ABSORBED LOCALLY and the crater is left visible from outer space. Herein lies the downfall of the "an asteroid hit it" scenario as far as larger events on Earth are concerned.

Let me repeat for the armies of "an asteroid hit it" aficionados out there. The impacts are absorbed locally and the impact is very visible. Additionally, although we have weather on Earth that will slowly erode to a slight degree, the impact crater will remain easily identifiable forever.

Almost immediately the claim of the dinosaurs being destroyed by a large asteroid or comet impact started to be questioned. Where was the crater that would be large enough to have been the impact to which they were alluding? It would have had to have been so large that it would be visible and easily identifiable.

In the search they found such a crater in the northern region of the Yucatan Peninsula jutting out into the Gulf of Mexico. Yes, Yes, Yes, this had to be it!

This ultimately led to the conclusion that was 180 degrees diametrically opposed to the "an asteroid hit it" conclusion. Further study showed that, in fact, the dinosaurs were on Earth at the time of this impact and that not even the dinosaurs locally were exterminated. Dinosaurs were definitely killed in the area due to the impact, but their remains continued to accumulate ON TOP of the region of the impact. So "an asteroid hit it" did not even kill of the local dinosaur population.

But, as with any after-the-fact news, almost no one was made aware of this little "fact." After all the history and discovery channel specials, after all the other glossy magazine specials, after all the newspaper hype that received blanket coverage around the world, the reality follow-on study that conclusively showed that the "an asteroid hit it" explanation did not hold water. It was not true. It never was true. Something much different had happened.

You can interchange the word "asteroid" with "comet" or even "minor planet" if you would like. The end result is that such a collision with Earth had been completely over estimated in its ability to destroy the planet. Certainly, there would have been devastation. But the fanfare that was boosted by "computer simulations" was no more than a lot of hot air.

There was a duo, William Napier and Victor Clube in their 1979 Nature article, "A Theory of Terrestrial Catastrophism" that suggested that when the Earth and our solar system crossed the galactic plane, this would disturb comets that were believed to originate in a region far from the sun, called the "Oort Cloud," and this would cause comets to come careening into the solar system wreaking havoc with Earth. Theories are so wonderful. They receive a lot of attention. The only problem was that, in spite of selling millions of books and receiving international attention, not a single aspect of this nonsense had any validity at all.

The idea that Earth and our solar system was somehow "crossing the galactic plane" was no more than a fairytale, imagined in the minds of these two and had no basis in reality. (In fact we see many YouTube internet wannabes today using the same kind of imaginary projections). We have no more idea where the galactic plane is as most of the galaxy is not visible and in fact there is no such thing as a thin plane. (The galaxy is a vast thick entity and we have no means of seeing, let along measuring, its extent.) Yet this was published in what is supposed to be a serious peer reviewed journal "Nature."

Secondly, there is the assumption that somehow something jarred comets lose from an imaginary thing called the "Oort Cloud." The Oort cloud was a concept originated in the 1950s in an attempt to explain where icy comets come from. Most long period comets have an elliptical orbit that seems to originate at a certain distance from the sun and also combined with the incorrect idea that comets are dirty snowballs, Oort projected the theoretical Oort Cloud. The only problem is it is so far away and so sparsely populated that no one could ever detect it. It is just that ... a theoretical construct needed to fill in the gaps of other failed astronomical theories.

Then there are the other aspects of the Clube and Napier fairytale. Would something disturb them and somehow many of these would end up smashing into Earth that is a speck of dust in the vast emptiness of outer space? All of this is befitting of an ostentatious (I should say Austentatious-like) play with a long list of improbably impossible characters. Yet it got press ... LOTS of press.

Today modern researchers are making the same claims without knowing that what they are claiming could not possibly be true and has been more than completely debunked in the truest sense of the scientific method. Experiment should rule supreme.

Other issues with the colliding asteroid theories are that it does not explain ALL OF THE DATA. Besides not being able to locate the craters that might be responsible for such activity, there are many other aspects that it does not resolve. Today there is a new burgeoning crowd of neo-archaeologists and neo-anthropologists who are taking up the catastrophism sword. They are aiming at the "Younger Dryas" period in which they think there was a mass extinction due to a colliding comet or asteroid. Since they cannot find the crater they conveniently (and without any evidence) state that the impact occurred at the north pole ice cap and sprayed the Earth with high velocity debris.

The real answer lies in large comets passing by

Earth that do not hit Earth at all. The meteor streams that accompany these comets include the streamers of diamonds created to the sunward side of comets (see my other texts dealing with these topics). And other fine metallic particles that are part of the meteor streams of comets, explains the peppering of regions where mastodons have been found with spray of micro particles. Passing large comets do not leave craters in the Earth's surface, but leave many other telltale effects, including pole shifts that result from the gravitational waves that wrench the Earth's surface.

The "an asteroid hit it" is not a viable explanation for any recent events (within the past 50,000 years) for the reasons given above. There is an effort under way in an attempt to replace the standard archaeology and anthropology with new catastrophic explanations, however if the people proposing these continue to offer explanations that could not possibly be true, then they will ultimately only strengthen the stance of standard ivory tower theories that they are trying to overturn.

The other issue is that these same people are trying to put dates on events, and this alone will doom them to pitiful failure. This is because we are just starting to crack the surface of this historically remote set of incidents. Every time you pick a date of an event based on certain criterion, ten objections immediately arise and the argument immediately turns to debunking the date. The heart of the matter never is addressed (that some kind of space-born catastrophe was responsible).

VII.
MISINFORMATION and CONTROLLED OPPOSITION

In the best of times there will always be more disinformation than real information. That is truer of the scientific world than the political world. An earlier chapter talked about the pyramid of knowledge and how it is all controlled from the astronomical departments (especially Cornell University where Carl Sagan set the stage for all time with his 1979 COSMOS series).

Most people do not realize that the players you see in the press or other news outlets are not the top tier of science in the world. The reality is that if you can see their faces and names, they are tier 2 scientists, at best. You will never see tier 1 scientists. This is what I call military science, since these are sequestered into positions and they have signed non-disclosure agreements that result in death or life imprisonment if they even make mention that they work for "the dark side."

Carl Sagan was the epitome of tier 2 science. The thing is that these people rarely know that they are just tier 2 scientists and that their entire role is to disinform the public. In my Education Program for K-12 and University Science Supplement I have about twenty lectures dedicated to Carl Sagan's COSMOS series of 1979 (and also the follow on series of Neal de Grasse Tyson who basically re-scripted the Sagan COSMOS series recently). The book COSMOS is the most read book in the world next to the Gutenberg Bible. The series is the most watched series in the history of the world. You would think that Harry Potter or some other program like Star Wars or Lord of the Rings would have been seen by more people, but NO, it was the 1979 Sagan series COSMOS.

Anything that is promoted to this level did not get there by accident or its own merits. This compendium of concepts was necessary to set the human mind into motion to believe that the universe was created in a big bang and all that you see is random. It also set the stage for the concept that MAN and his history are well known and the result of EVOLUTION. It implied that we know all about this. Below is an excerpt from the explanation of my education package where I stress the impact on the public (especially children). Note that the message is that you evolved and are just a little bit smarter monkey than the ones still out there in the jungle. Look at the implications of such a doctrine. Why was Sagan's message broadcast and promoted to the far corners of the Earth to the point where literally 50% of all people on Earth have read or seen this information?

> ***The Professor's Course replaces the atheist communist science math education forced on the youth of today!***
>
> ***Did you know that "accredited" science and math education has been deliberately created to remove logic and reasoning to make the population controllable by the elite?***

The "accredited" public school education programs of today (also required in private schools) were designed to create an atheist communist state. Don't let this happen to your children. A secret government document titled "Silent Weapons for Quiet War" (1954) outlines the purge of real science and mathematics from schools. This was solidified in 1979 by their "movie star astronomer" Carl Sagan, who immortalized the communist atheist agenda in his "COSMOS" series and book. This set the material basis for all future education of the "masses." Accredited Science and Math courses are designed to "dumb down" your students and intentionally remove LOGIC and REASONING. Up is down and down is up. Little boys are told they are little girls and little girls are taught that they are little boys.

In his course set, the Professor first identifies the travesty of "accredited" science and math education with its set of unrelated "facts" (which students are forced to repeat). With Professor McCanney's 13 books, numerous CDs & DVDs, other supplemental materials, and 180+ one hour MP3 lectures, the Professor teaches science – reason – Logic and replaces the atheistic communist agenda. It is available in a "Basic Package" and "Everything Package" Course sets. The "Everything Package" course set includes

an additional 89 one-hour MP3 lectures and supplemental materials INCLUDING a 20 lecture series breaking down the Sagan "COSMOS" series propaganda, replacing it with sound science and reason. Sagan's subliminal message was that God did not create the universal design, but it all happened by itself one day in a "Big Bang." He also taught that you are all "evolved" and are 'just a little smarter monkey than the ones still out there in the jungle'. It took away your God given soul, your identity, your birthright as a free human being AND YOUR GOD CREATOR and replaced them with voodoo explanations (and you thought Sagan was teaching "astronomy"). Professor McCanney's course is essential for all humans but especially Christians. We are all created by the same creator. With this renewed knowledge you can reclaim your rightful heritage as a human being. Grandparents and parents, this course is essential for your children. We first educate the parents, then you educate your children.

Modern education is designed to destroy your spirit. Most of all, at every turn, the Family is under attack. When you hear educators use the words "diversity" and "inclusive," it really means "teach all the kids to be the same" and "be ashamed of yourself." We have to teach our youth (starting at a very young age) that

they have a spirit which makes them unique AND which tells them right from wrong … that they as an individual are important. It is this spirit, acting with their physical body, that sets them apart from the animals who only live by reacting to their environment. This is what gives man his creativity and uniqueness as an individual. It is what make us children of God."

All science must rigidly follow Sagan's set of theoretical principles from the top level of scientists to the most remote kindergarten teacher. But there is a more sinister part of misinformation. Long ago in think tanks associated with the CIA and other "intelligence" agencies, and going back to the Middle Ages and well into the deepest darkest corners of ancient man, a technique has been in use that caps off the control of information to the common man. Misinformation is one thing. Direct … In your face … But the more subtle aspect of misinformation is to control both sides of the dialogue. One creates controlled opposition to make the "authorities" look right and indisputably correct.

To locate the opposition is easy. You look for people or groups that are eager and unsuspecting and with large egos that are easily stroked. You give them fame and feed them support under the table. Look for people with lavish web sites, travel budgets, etc., and no visible form of support. They have handlers who guide their message and make it within the limits of what they are willing to let the public see. It is generally filled with things that could not possibly be true and thus blocking any real opposition from being seen by the masses. Today this is in the form of internet web pages, Netflix or history channel specials, and interviews on top level radio or TV shows.

The entire purpose is to prevent real opposing ideas from ever seeing the light of day. The other purpose is so that the reigning authority figures (the Carl Sagans and Neal de Grasse Tysons of the world—what we call the "White Hats") can ride in and save the day for standard knowledge.

Misinformation is an art form; it has certain aspects that are always used, but the most important aspect of this is the universal use of controlled opposition. That is why for example on YouTube you will never see my name and if anyone ever puts a blog note referring to me or my work, it will be removed immediately (although the person who put it there will see it and think it is still there, no one else will be able to see it). This is all going on while people steal my work and post in YouTube and additionally turn it into crazy lunatic topics, they are never censored at all.

But one has to ask why this is so important to the people who control the news media and ultimately the control of public perceptions? The reason is simple ... CONTROL. It took me a long time to see the connection between scientific or mathematical constructs that I have produced, and the control of public perceptions. If people looked up into the sky and saw a raging comet as was discussed in ancient writings, they would immediately know that the leaders of the governments of the world, the banking communities, the insurance industries, and on and on—all of these—would have no power over that thing out there.

They would immediately lose control that they have worked so hard to secure. The evening news is filled with what we call "fear porn." The stock market, housing crisis, recession, political figures, corruption, banks failing, war, and the list goes on it is endless. All of this gives the false impression that there is somewhere a leader that is going to solve all of our problems and we need all of the leaders. The one thing that would send this all packing in a New York second would be the idea that there is something out there over which these clowns have no control.

It is the Wizard of Oz syndrome. "Don't pay any attention to that man behind the curtain." A space shuttle blows up with a teacher on board. The message to all youth "you cannot go into space – danger – only NASA and its teams of engineers allowed."

VIII. ARCHAEOLOGY ANTHROPOLOGY GEOLOGY

If you have ever seen what happens when someone becomes a thorn in the side of the "intelligence community" the end result is always the same. To put it bluntly ... they "muddy 'em up." There are many ways this is accomplished and this, too, is why there is such allegiance to these psycho agencies. The people on the inside have seen repeatedly what happens to someone going public and ratting on the operations that are "top secret." They are top secret more often than not because if the public ever found out about the illegal activities that were going on, they would certainly rout out the corruption. But the perpetrators continue with the veil of "truth, god and the American way." But notice god is spelled with a small g here because their god is not the same as your God.

The fields of science are all kept in line with a simple hierarchy that will ruin the career of the offender that strays outside the hallowed halls of academic rigor. When Sagan set up the COSMOS series it was meant to finally set the regimented rules of the road for Astronomy. The end result was that all disciplines had to follow the same norms. Imagine someone discovering that man did not originate on planet Earth and was a star child that traveled a great distance to populate Earth. This, of course, would automatically fall into the realm of disbelief by the general public because they have been conditioned to believe that man is a product of evolution (we descended from the apes remember?).

What about other topics like a pole shift on Earth. This can only be the topics of lunatics and crazy ladies that talk to aliens. Do you see the pattern here? The prior chapter talked about misinformation and controlled opposition. These are cases in which the public has been primed. How often do you hear someone say “I just don’t believe that”? My question always comes back “Based on what?”

In general, it is based on psychological conditioning that makes people think that something would be out of the current realm of possibility because of their conditioning. Yet they are convinced that they are correct. Conversation over … end of discussion. If you continue talking, you will only get a glazed over deer-in-the-headlights look and “fact checkers” will quickly jump on it to dispel it (and who would not believe the self appointed fact checkers that populate the internet today?).

Pages like Wikipedia follow the party line. The self-proclaimed truth editors come from all areas of “standard western knowledge” and bash anyone and anything that is not of the fold. For posterity I am pasting here the posting on Wikipedia regarding my own persona. May I remind you that Wikipedia has topics down to the most minuscule topics known to man? Once there was a Wikipedia page that was extremely offensive to my name and work but it only sufficed to send people to my web page where they ended up becoming fans of mine and buying my books. I wrote to Wikipedia thanking them for all the traffic they were sending to my page and to "keep up the good work." Today here is the post listing for James McCanney (mouse copied from Wikipedia).

The Page *James McCanney* does not exist

According to Wikipedia I do not exist. Given the long list of public engagements and publicly available work that I have produced, including peer reviewed papers in astrophysics journals, and given my long list of appearances on radio and television, both in the USA and internationally, it is hard to imagine such a situation. However, there is another web page called "rationalwiki" that posts the following short quip:

> **James McCanney is an astronomy crank and a conspiracy theorist. He's chiefly known for his "electric comets" claims or, as he dubs it, "The Plasma Discharge Comet Model," but he's also a fair example of crank magnetism and appears to believe in all kind of fringe ideas. In addition to pseudoscience, his website flogs survivalist supplies. He has an awful podcast, where he drones on and on and on about his kookbooks. The podcast occasionally appears on shortwave radio.**

Whoever wrote this has certainly never spent more than a few seconds trying to look at my work and certainly never met me, and even more certainly does not have any intention of pursuing it any deeper than this cursory diatribe. He totally misrepresents anything that I have ever presented, apparently not agreeing with the "editors" belief system. My past is extremely rich and I have major discoveries and works in many diverse fields. I am recognized worldwide as a Tier 1 scientist and am a regular expert source on international radio and television. The fact that this "editor" resorts to bizarre personal attacks and verbiage only makes him look like the lunatic.

Remember this is not so much an attack on me and my work, ***but a message to anyone in academia or the other branches of "standard western knowledge" to not touch me or my work with a ten foot pole. That is the true message intent.***

The other subliminal message is that "do not even think about crossing the line to talk about these topics." It is off limits and your hard-earned career will be cut short if you even mention affinity to the ideas of McCanney. This keeps the ranks in line. The thought of losing tenure or hard earned funding publication rights is more than any university or government agency employee would be able to withstand. After all, where is a PhD in any of these disciplines going to get a job in the real world? Their paid scientific vacation, retirement, and entire career would be lost in a flash. That is the true message.

This also gives anyone the green light to use this as an excuse to ignore any of my messages and in fact that it would be open season to steal my work unabated (as long as it is twisted and kept within the realm of standard science). This I call "the creeping crud" as my work is constantly being infringed upon and pilfered in the same academic circles as the only path to correct answers is via my work that completely replaces standard science.

In summary, the physical sciences all come under the umbrella of the astronomical world and must always adhere to these "standards."

IX.

HOW LONG DO ARTIFACTS LAST?
With Discussion of Physical Pole Shifts

One of the greatest difficulties in analyzing the past is the short time that artifacts last in the ever changing environment. Take for example my discovery that the Nazca Lines constitute a fractal antenna array (see my white paper "**The Nazca and Palpa Lines – Mystery Solved**"). One of the quagmires is that these are clearly designed for interstellar telecommunications. But as we see them today, and as they were very first recorded in history by the Spanish, there was never any material in the tracks of the figures that would have been used to transmit electro-magnetic signals.

Our estimates are that the figurines are at least 2000 years old and could easily have originated as much as 10,000 years ago or more. The dilemma occurs when one realizes that anything setting in these tracks from that long ago would simply be gone with the erosion of time. If we took a brand new automobile today and set it out on the Nazca desert, what would be left after a period of time? After one year the paint would be fading and the rubber tires would probably be flat.

Within ten years there would be paint faded and, in some places, removed with the steel oxidizing. In twenty years, the paint would be flaking off and areas of oxidation would be apparent. The interior would be coming apart and plastic parts would be changing shapes to the point they no longer fit with the metal parts. In 100 years, most of the car would be gone. Within 1,000 years there would be very little left, except for the heaviest of metal parts, such as the engine block and frame. After 2,000 years there would not be anything recognizable and after 10,000 years there literally would be no trace of the car at all.

Now imagine that the technically advanced artist cultures who created these lines put iron in them that was mined or smelted from the iron rich rock that is prevalent all over the Nazca planes. Imagine that they paved the trenches with iron (or other metal). Literally none of that would remain today. Imagine that they built atmospheric electricity generators (essentially Tesla Towers like at Wardenclyffe) on the mounds that connect all of the antennas into a circuit board-like array to power the electrical signaling. None of this would be there today after thousands of years.

Now imagine that there have been at least five major celestial events in the past 10,000 years on planet Earth, including a massive flood that destroyed nearly everything and raised the sea levels between 200 and 600 feet. Imagine that there was another event so devastating that the poles shifted by about 30 degrees and there was major mountain building and another near mass extinction on Earth.

Imagine further that entire populations (Atlantis) sank into the ocean in a single night and everything we know about it is only found in brief myths from ancient priest cults. This is where we are today on planet Earth. In addition to trying to determine what really happened and when, we are beset with a completely incompetent and in control university, museum, and government system that does not allow any real research to occur.

Imagine that there are controlled opposition independent "scholars" that have been promoted under the table by the same controllers that want the public to remain completely incognizant of the true history of the world and who are attempting to reduce all of this to a single event that they claim happened in the far distant past, long before any of the real events actually took place.

Imagine that data, which is scarce and difficult to accurately date, is used to support the misinformation with the end result that the public is kept believing the … "Oh, this event took place long ago and does not and will not affect me." The wizard behind the curtain just maintained the cloak that hides the truth from the unsuspecting public. One of the other main issues is that during these explosively damaging events, the Earth's surface was changed and literally covered by large amounts of new material. This leaves the majority of real evidence completely covered up and will never be uncovered. I believe that if we were to dig hundreds of feet under solid ground, that we would find more lost civilizations of which we have no record. This has occurred and is ignored by modern "science."

The ability to find artifacts and their lifetimes after major disruptive events makes it hard to interpret and even harder to put into perspective. I am adamant that before dates are ascribed to ancient cities and civilizations, that we have very good reasons for this dating. If not, we are left with an improbable history just like we see today. This gives the controlling few exactly what they need to keep the public thinking about the evening news.

There are so many effects caused by large passing comets that are explained in prior books of this series that we cannot begin to cover them all here. The serious readers must obtain the full library of books, CDs, DVDs, lecture series, etc. and study it. The risk in offering superficial partial explanations in this book should be obvious. It would only fall prey to angry defensive astronomers, archaeologists, geologists and others who would wish to disregard and dismiss out of hand. So, I require that serious readers must take the serious time to read the past books, listen to the CDs and hundreds of lectures, and watch the selected DVDs.

The one topic that I will discuss briefly is the concept of physical "Pole Shift," because of the massive amount of work that has been offered by independent researchers in this field of study including the large body of recently released conclusive supporting information.

There are two kinds of pole shift. Many times per day our magnetic field experiences variations due to electrical currents passing by earth from solar activity and our discharge of the solar capacitor. Occasionally, it temporarily reverses 180 degrees. The majority of the magnetic field is based on my five layer model as part of my work with solar system electromagnetic effects. This should not be confused with the rare physical pole shifts that only can be caused by a gravitational wave of a large massive planetary-sized comet nucleus passing near to Earth.

In this case, the land is stretched into a tsunami-like wave that is wrenched over the Earth's inner core (which continues rotating as before). A detailed drawing with physical explanation is included in my "**Surviving Planet X Passage**" book. This explains plate movements, mountain building, water tsunamis, movements of the oceans, and a myriad of other effects (all occurring at the same time along with other effects of the large passing comet). This is the most likely scenario for the parting of the Red Sea as Israelites crossed a land bridge on their escape from Pharaoh's Egypt with Moses. The sea washed one direction giving them time to cross the narrow land bridge and shortly after washed back, killing Pharaoh's army. The chariots have been found in the Red Sea. These are real stories, not fiction.

When I was a young man (early 1970s) I lived in Mexico City and spent much of my time traveling and researching the hundreds of Mayan, Olmec, Toltec, Aztec and other ruins that dotted that marvelous country. I should write a separate book just on my travels and adventures, as Mexico and Central America were truly the wild west. At the time I was just observing. Later in my life all of this would come into play as I realized the significance of the Myths Magic and Mayhem conveyed by these past cultures.

I spent many days on archaeological digs with world famous archaeologists as ruins were uncovered for the first time in thousands of years. I always found it interesting that they removed the dirt and dumped it as unwanted refuge in some remote area. To me, this should have been studied to determine its contents, since there should not have been any reason why a thousand year old structure would be completely buried. We are talking about entire massive pyramids and not some small relic in the Yucatan jungle. As the subway was being built, they constantly were unearthing ancient ruins buried deep below what had been occupied by the modern Aztecs.

I was fascinated by the drawings of Catherwood showing destruction of Mayan ruins as they found them in the 1800s. I spent time in the Mexico City Museum. It was nearly twenty years later that I returned with surveying equipment to the major sites of Chichen Itza, Uxmal, Tulum and so many others, to determine the layout of these remarkable cities.

What I discovered was that they lay far from the cardinal directions of today's Earth. Chichen Itza's main structures, including the main pyramid were 13.5 degrees off of true north. Oddly enough these were all dedicated to the comet Venus (not planet Venus, but comet Venus) who they worshiped as Quetzalcoatl the Plumed Serpent God of the night sky, whose heart eventually became the planet Venus. Venus as a large comet (not the sun nor moon nor any other celestial deity) was central to the Mayan Popul Vuh. Venus today is just a small planet between us and the Sun. The reality is that Venus was a large raging comet that encountered the Earth on two major occasions and eventually (because of the comet "Tail Drag") its orbit was circularized to become the hot steaming planet we see today (exactly as depicted by Immanuel Velikovsky in his epic books).

I then used my exact surveys with maps to triangulate the location of the old north pole, now that I had accurate maps and surveys of these sites in hand. What I discovered was that there definitely had been a large pole shift and that the old north pole lay in the province of Ontario, Canada, north of the USA state of Michigan. The best location put the old north pole in what is now Hudson's Bay.

At this point in my life, I not only had my well developed celestial mechanics explaining how large comets would become planets, but had as well my long standing publications explaining every other aspect of the solar system and galactic evolution. Part of this body of work explained how large passing comets would cause major pole shifts in the existing planets among hundreds of other effects.

I had completely re-written all of astrophysics and astronomy showing that the standard theories could not possibly be true. I had proof on the ground of a massive pole shift that matched with so many other unexplained facts, such as temperate climate mastodons flash frozen in the Siberian tundra with undigested plants in their throats. They were discovered standing on all four legs (they died so quickly and the snow hit so fast that they did not even have time to fall over).

Thousands of years later they were uncovered and used as food for the trans-Siberian rail workers and sent to Paris as a delicacy. They were dry frozen so quickly that they did not have time to decompose. The delicate flowers in their throats' digestive juices would begin to break down in five minutes or less, but were fresh as a daisy thousands of years later.

To investigate further I visited a bison (buffalo) farm in Wisconsin and talked to the resident manager. I asked him how long it would take to freeze a whole buffalo. He looked at me like I was a little crazy and said it would take weeks. Then I asked him how long if the buffalo was alive. He looked at me like I was really crazy and simply stated that you could not do it. The animal has so much mass and internal heat that it would simply not be possible. Recently an entire frozen mastodon was extracted from the Siberian tundra. It was so heavy that a special helicopter from the USA was employed to lift it. Modern science, in its never ending quest to make you believe that all of this happened slowly, stated that continental drift was responsible. How absolutely absurd as this would take millions of years by their own standards.

So how could mastodons, which are ten times larger than a modern day bison, become flash frozen in minutes while still alive standing on all four feet? At this point, lets return to the other side of planet earth where the Mayan ancestors, the Olmecs, populated the area at the time of the great pole shift. Their stone cities were constructed over the landscape. We do not have an accurate time frame for the Olmecs nor their northern neighbors the Toltecs. The Mayans we have record of today thrived in what is now Yucatan, Mexico, into Belize and Honduras around 800 AD. They rebuilt the destroyed Olmec cities using the same grid work as used by the Olmec, not changing to the newer current celestial coordinates. In their Myths Magic and collective recollection of the Mayhem wrought by the great comet Venus (Quetzalcoatl) they covered their old pyramids and structures every 54 years in thanks that Venus did not destroy their world. It turns out that Venus, the large comet, approached Earth on two occasions about 54 years apart with the second apparition being the closer and more devastating. Their descendants, the Lacandones, were still there when I was a young man visiting in the early 1970s.

I have a story recounted in my other books worth noting here. I was camping with a friend in the jungle off of Tulum in 1971 (this was long before current day tourism and "Cancun"). A new road made from sea shells had just connected south to Tulum and we were literally the first people to traverse that road (it can still be seen running next to the current highway from Cancun to Tulum). We were living off of coconuts, conch or fish from diving on the reef every day, and rice.

I heard of a Lacandon settlement off in the Tulum jungle, so one day I set off to find it. I eventually arrived to be greeted by local men who invited me eat lunch with their family. They were short and small boned people with the tallest being about four feet. Some of the women were slight of build and looked like they had bone structure of a small bird. I ate lunch with them, and late in the afternoon I was about to leave when I saw a small hut with a woman attending what appeared to be a store.

I entered to buy some fruit for the return trip. She did not speak Spanish and I did not speak Lacandon dialect, which I am told, sounded like the original

Mayan. The only modern product there was canned condensed sweetened milk. Above her head, sitting on a perch, was a Quetzal Bird (bird of paradise that roosts in the day and flies at night). As I looked up, the bird left its roost for its nightly flight and flew directly over my head. To say the least, I got a bird's eye view of the plumage on the under side of this magnificent bird with its long flowing tail. The plumage was precisely that of a long multi-tailed flowing comet with the coma (the under breast and wing area) flaring back and the long sunward spike. Little did I know that decades later I would remember this as I learned about Quetzalcoatl, the plumed serpent god of the night sky that represented the great comet that eventually become the Planet Venus.

The Maya, as with all pre-Columbian civilizations, rebuilt their cities every 54 years after a Passover-like event for which they went into hiding for fear that the comet Venus would return and destroy their world. (This celebration had been occurring for thousands of years, just as modern cultures continue to celebrate the Passover of the Israelites before leaving Egypt with Moses). The Passover event has been replicated by Native American tribes, as well as other places in the world. For further information listen to my 18 hours CD MP3 lectures covering these topics in detail. I cover the many plagues that beset the Pharaoh and Egypt, both before and during the passage of the comet Venus. The Passover literally represents the ancients sealing themselves into small huts to avoid the poisoning taking place from the massive passing comet. If such an event were to take place today from a large comet (not the small comet such as Halley's comet), the only survivors would be those who knew how to seal themselves off from the poisonous gases and pollution from the passing comet.

At this point in the discussion of physical Pole Shifts, suffice it to say that I have discovered at least two pole shifts based on physical measurements of archaeological ruins. One was about five degrees, while the last major pole shift was about thirty degrees from the old north pole location in Hudson's Bay to the current north pole. You will find that our geomagnetic field is slowly becoming aligned with the physical north pole having originated in the old north pole region of Hudson's Bay.

Another data point turns up in the Nazca Palpa Lines structures. One of the earliest purely fractal antenna designs from the older Palpa set of antennas is one that points perfectly aligned with the old Hudson Bay north pole. This indicates that some of the lines were built prior to the last pole shift while some were built after.

I have to make the point again very clearly ... I do not place dates on any of the events since it is, in my opinion, too early to determine exact dating. The reason I am emphasizing the physical pole shift, which is just one of the myriad of effects of large passing comets, is because there has recently been a flurry of activity among scholars showing exactly the same result. This reverts back to what some may consider as "discredited" research of earlier scholars. Fortunately, modern researchers are breaking rank with old archaeological ideas. This has to include the fact that celestial events have seriously shaped the development of Man and Earth. Once again, the idea that a small asteroid or comet hit Earth does not explain what we see, and even if you were to still imagine such, there are no craters that would be large enough to begin to move the entire surface of earth. There is nothing about the "an asteroid hit it" scenario that holds water. The only explanation is that of a large planet, in the form of a Plasma Discharge Comet passing near enough to earth with a large gravitational wave that would cause the physical pole shift.

There is no longer any question that there have been multiple significant physical pole shifts of Earth's surface in both recent and far distant past time frames with the possibility that at least one major event occurred as recent as 1500 BC.

The only reference in this book, other than to my own prior writings and lectures (only available on my own web site www.jmccsci.com), is to the great work of one whom I consider to be an associate scholar on this topic—Mark Gaffney. Mark has examined many variations of geology, ancient architectural alignments, marine fossils and ocean records and more, demonstrating the fact of physical pole shifts. Some of these lend themselves to dating techniques that give us an idea of timing of the events. Mark, too comes to the conclusion that the last old north pole location is located in what is now Hudson's Bay. For further reading see his book **"Deep History and the Ages of Man" by Mark Gaffney.**

X.
DID MAN EVOLVE FROM THE APES?

One of the greatest deceptions in all of history is the Darwinian theory of evolution and origin of the species. Like many textbook topics, it sounds great on paper. When all other topics use it as a starting point, circular reasoning comes into play and very few people are logically connected enough to realize it (including university professors). There are many cases today of circular reasoning. The Big Bang, General Relativity, Dark Matter, Dark Energy, the origin of the sun and planets and solar system 4.5 billion years ago, plate tectonics, the Oort Cloud, and the list could fill volumes.

Primarily, all of these theories have one thing in common. They were proposed to solve a nagging question in our never-ending search for the origins of what we see. All of them also contain a leap of faith for which there is no scientific experimental evidence. In all of these cases, there is only secondary circumstantial evidence that has led investigators to believe that they are true. In none of these cases can you get the problem simplified in a laboratory so that definitive experiments can be performed. Yet, they are all believed by the vast majority of people to be true.

Once they are repeated enough times it is set into the human psyche to the point where stating the obvious, that there is no experimental evidence to support any of these, becomes heresy. Let us examine the leap of faith in Darwin's Theory of evolution. It requires something that has never been found called "the missing link." What Darwin actually identified was the apparent slight changes of various animals that occur over time. But just look at the human race. Every generation could be viewed as different from the one before. The entire problem with Darwin's conclusion is that he, too, had made the assumption that all we see had to happen on Earth. This potentially could be a valid assumption, but it is an assumption for which he (and even science today) has no real data or answers.

The leap of faith that Darwin made was that small superficial changes could be extended across what were called the "species." Could the ape have descended from man instead of visa versa? That would actually make more sense since less complex things usually evolve from more complex things (if you look at entropy for example). But without evidence, neither path could be proven. Today evolution is a cornerstone of astronomical theory, and was an essential point made by Sagan in outlining his history of the world.

The theme that the world did all that we see by itself, and no major outside effects were possible, is central to Sagan's theme and all of science today. Sagan had two evolutionary themes that he cast into stone. One was the physical evolution of the cosmos (Big Bang) and the Darwinian evolution of biological life. BOTH negate the idea of a God or grand design. Both designate that the universe must be finite, which is more a figment of man's finite imagination (man cannot imagine the infinite).

Darwin made an assumption that the small

changes seen within what he called "the species" would extrapolate to large changes over enough time to become separate distinct species. This assumption is false. Thus, we witness the never ending search for the "missing link," which, to date, has never been found. Imagine that, in spite of millions of years of "evolution" and all the tens of thousands of intermediate stages that would have had to have "evolved" between ape and man ... they have never found a single one.

It is like the search for Dark Matter and Dark Energy. It is theorized because the Big Bang theory does not make sense, so they create more and more nonsensical theories to put band-aids on the failed original theory. The history of this is complex, but follows the same "reasoning" as Darwin's evolution.

The concept of the expanding universe is one of the greatest examples of circular reasoning that was ever concocted by the mind of man. Edwin Hubble was the first to have good equipment, using spectrometers to observe galaxies and their red shifts. All were showing red shifts, so he proposed the expanding universe theory. It gained traction and so all data was interpreted to "support" this concept.

The only problem was that his original assumption was that the red shift was due to things moving away from us. That assumption turns out to be incorrect. The real cause of the red shift I discovered, when I was a faculty member at Cornell University in 1979, called the Induced Electric Dipole Red Shift. It is a red shift caused by photons moving out in the non-uniform electric fields that surround fusion based objects. Once again, circular reasoning within inner circles of academia has solidified a totally bogus concept. Scientists boast mountains of data; however, all the data is interpreted based on the assumption that the universe is expanding. This is circular reasoning.

If you took a logic course at any level, the very first thing you learn would be logical tautologies, and the very first of these would be that if you assume something, and then base other statements on the initial assumption, then all your other statements are only as good as your original assumption. In no case can the statements that follow be used to prove the initial assumption. This is like the perpetual motion machine of logic; it simply does not hold water. But it is used all the time in modern "science."

Let me give another few examples. The Big Bang and galaxy arm models depend on the assumption that gravity is the only large scale force in the universe (and essentially this is correct). The problem arises when one adds the concept that the universe began in a Big Bang original explosion and that everything somehow evolved after that, and secondarily that it all happened near the beginning of time. Since this leads to some very odd conclusions, theorists are required to invent things that they have never been able to find, let alone define. Thus, Dark Matter and Dark energy are claimed to make up almost all of the known universe, in spite of the fact that no one can tell you even a single property of the Dark Duo.

These are the "missing links" of Astro-physics. Just like Darwin's evolution and man allegedly evolving from the apes, what should be everywhere and in plenitude, staring you in the face at every turn, there is nothing. The same holds true for the illusive Oort cloud of comets that can never be found or discovered. It is the same for Plate Tectonics where no one can tell you what herculean forces push entire continental plates sideways around the globe. Now one last example ... General Relativity.

Albert Einstein objected to the General Theory. Although his Special Relativity has been more than laboratory verified, General Relativity is based on an assumption that has never been laboratory verified in a repeatable experiment. Mountains of data have been taken, assuming that the General Theory is correct, but as with real science and logic, an unaffirmed assumption does not a confirmation make. The crux of the matter is that Einstein made what is called a Gedanken Experiment or thought experiment. It is also called the Principle of Equivalence or the famous elevator analogy.

Essentially, one mentally compares an elevator setting stationary in Earth's gravitational field with the same elevator accelerating in free space at the same acceleration as on Earth. All experiments should result in the same results. There is a leap of faith ... a missing link. The assumption is that electromagnetic radiation should act in a gravitational field as any other gravitationally responding entity. As it turns out, much to the chagrin of astronomers, this assumption is incorrect. Einstein worked on this problem until he died. (See my other work for a full explanation of this dilemma.)

One of the fundamental assumptions of all of modern understanding of MAN and Earth is that all of what you see must have come directly from Earth itself and with no outside forces or influence, apart from a small comet or asteroid impact. In all scenarios, Darwinian slow evolution is the only possible answer. In this scenario the only possible place for man to come from (outside of direct creation by an almighty creator) is ... evolution. It all makes so much sense. There are some nagging questions that arise even in the evolution only representation. Man is different from all the other animals. He has a certain "spirit" that makes him different. Is it, as Carl Sagan explains in his atheistic Darwinian dialectic, that man's brain is just a little bit bigger than our uncle apes still out there in the jungle, or is there something else going on here?

XI.
DOES MAN HAVE A SPIRIT?

Where things start to get murky in the waters of evolution is when, all of a sudden, living things start to emerge from what allegedly before was just inert matter (rocks, gases, liquids, plasmas, fiery galaxies with billions of stars, molten forming planets, and last but not least ... little snowy comets). How did the protons on the far side of the universe know to be exactly (I mean EXACTLY) the same as protons on our side of the universe, if it all happened at random with no grand design or creator?

Why does the carbon atom act differently in a diamond lattice than it does in the molecular chain of a whale in the ocean? "Science" cannot answer that question. Why did our allegedly near relative ancestors, the apes, not develop quantum mechanics and telescopes to view the stars? In fact, do apes even know the stars are there? Do apes look up and make up names for constellations and try to figure out what makes stars bright?

If you have half a brain as a scientist, you have to realize that we are only discovering what was already there long before man was able to conceive of the questions to ask regarding how the universe works. Some scientists think we know about all the workings of the cosmos. Carl Sagan sure came off as though he thought as much. He was very sure of what he was saying.

All the great scientists are, at some point, claimed to be "atheists." You regularly see articles stating that science does not need God. Only a pompous self-absorbed moron would imagine that the organization of the universe was just by chance and that there was no plan.

I have a belief that God has a plan for atheist who believe that after you die you have no spirit that continues on. That plan is to grant them their belief and eliminate their soul from existing. The reality is that the great scientists mentioned above did not belong to organized religions. So in the 1950s sense of the term "atheist," they were non-secular and therefore atheists. In reality they were not atheists. In many respects they were more believers in a divine creator than many religious people, since they had an understanding of the amazing complexity, organization, and consistency of the universe of which they were struggling with.

The idea that man has a spirit is easily verifiable. I have personally gotten to know and interact with my inner spirit, although this was never any part of my education nor did any one teach me how to do this. There are many things that we cannot ascertain from the spiritual universe. I liken it to a whale in the ocean that is able to occasionally breach and see the world outside of its water world. For that brief second the whale has a glimpse of an area that he cannot comprehend. The same is true of man and the spirit world.

The idea of spirit is manifested, for example, in our expression of music and listening to the music of artists. Some music has universal affects, such as happiness, sadness, melancholy, patriotism, and strength ... to name a few. This is the spirit interacting with the music and our bodies. Science is always trying to figure out how the human brain works and even projecting that artificial intelligence (AI) will be human like and take over all of our functions. The thing this is missing is the spirit that lives in every human being.

The spirit is what gives man his innate sense of right and wrong. Do you notice that criminals and con men all know they are doing wrong and mask themselves as being upright normal people? If they had no sense of right and wrong, they would not bother with any of this. It is universal, down to the most corrupt and baseless of humans. This gives rise to what we call "natural law" and "inalienable rights." These are rights given by the divine creator to man. This goes along with the idea that man has a free will and is the only creature on Earth that has the ability to create his own future and fate. But with freedom of choice comes responsibility.

If you do not believe that man has a spirit then you are missing one of the greatest marvels of being human. If you think you are, as Carl Sagan wants you to believe, just a little bit smarter monkey than the ones still out there in the jungle, then I feel sorry for you who are missing out on this miracle of being human.

Where and when did man become infused with a spirit? This is part of creation. Are there other intelligent beings out there that have spirits and with whom we have a common heritage? Are we connected to spiritual signaling from the sun and other objects in the universe? If modern science cannot put a meter on it and measure it, then to them it does not exist.

The first test of having a spirit is to test your free will. Put your hand on the table or desk in front of you. Make a conscious decision to raise your hand on the count of three. As you get almost to the count of three, you change your mind and decide, at that moment, to raise your hand at the count of ten. You just interacted for the first time with your free will.

This is also your first step at realizing you have a spirit living inside of your body that has an energy field and is also connected to the outside universe. Anyone attempting to understand man and his thought processes without including the soul (spirit) will never understand the true meaning of man. To understand the true and full history of Man and Earth where we live, is not possible because we were not there when any of this was put into place. Maybe it has existed forever and will continue to exist forever. We have no way of knowing.

In my education program for K-12 University students and their parents, the first thing I cover is how to teach young people that they are important and that they are unique. No one else is like them. They are special and have special talents that no one else has. Their job in life is to pursue their life and raise them as adults. Young people who are treated like adults will act like adults even at a very young age.

Man's birth right is in the stars. One will find the corrupt atheist controllers, that attempt to sequester nations and people, are psychopaths. Their first goal is to strip people of their individuality. The public school systems entire purpose is to accomplish this goal. They create adults that have no idea who they are or what they should be doing with their lives. Understanding that man has a spirit, and in fact a collective spirit, is fundamental to understanding the true history of man.

XII.
IS THE SUN'S ENERGY CONSTANT?

The history of Earth and present conditions on Earth are dependent on a second by second input from the sun. This should not be a startling affirmation. What may be startling to most people is that it also changes on a second by second basis. The scientific theory that states that the sun is constant and has been the same for 4.5 billion years is complete nonsense. There is no scientific evidence for this. As with most non-nonsensical theories, scientists had to come up with a date to make themselves look like they know what is going on. Partial to this belief is the concept that the sun's energy comes from its core, by way of nuclear fusion.

The reality is that the fusion of the sun takes place in its atmosphere. This was my theoretical work from my "Cornell Years" and was verified by a simple but elegant experiment conducted by the atmospheric physicists I was working with at the University of Novosibirsk in central Russia in the mid-1990s. It involved an x-ray detector that they scanned across the surface of the sun and, by subtracting off the surface layer x-rays, were able to confirm that the solar heating is higher in the upper atmosphere and that the inner core was cooler than the atmosphere.

This proved that the sun's fusion is in its

atmosphere and not in its core. This also meant that the solar fusion base atom hydrogen had to be replenished on an ongoing basis. These same physicists saw that the sun's energy would wax and wane based on the input availability of hydrogen coming in from the galactic environment. The sun's energy can and does change on a very short time frame.

NASA built a probe to study what is known as the "solar constant" or the amount of energy coming from the sun at any given time and in given wavelengths of the electromagnetic spectrum. They set their timers to months and expected little to no change over long periods of time. The data came in showing exactly what I am telling you here. They physically measured changes down to one second, which was the limit of their equipment to measure change.

To understand Earth and the history of man, it is important to understand all the relationships of Earth's climate and our solar environment. Todays more than bizarre "climate science," which myopically only looks at the belief that CO2 would affect the climate, is so misaligned with reality that it is not funny. It always was just a politically motivated control mechanism (primarily population control and control of the world's water supplies).

During the passage of comet Hale Bopp, the solar output flared and Earth experienced x-class flares on a daily basis. This is because Hale Bopp was discharging the solar capacitor on a daily basis. When Hale Bopp left the solar system, the sun went into a long quiescent period that it is still recovering from today.

In January 2022 a large south pacific volcano erupted sending for the first time in recorded history a plume of ash into the stratosphere almost to the limit of outer space. That plume is currently covering the Earth and is causing a global cooling that may last years. The climate alarmists who incorrectly only consider one factor (CO2) chose to ignore this since they make a lot of money off of the support of this less than scientific claim. The idea that there is "universal acceptance of the science" is absurd. When is the last time you saw scientists agree on anything? The reason they agree is because someone who wanted climate change to be used as a point of control threw a lot of money at the universities for "research." In today's science world, grants and funding drive the results, not true science. They monopolize their field hungry for the grant moneys.

The final point regarding the sun as part of my theoretical discoveries is that the majority of energy entering and affecting our weather systems is not from the light but from electrical currents as we discharge the solar capacitor. We interact electrically with other celestial objects continually as do all of the other planets, moons, and smaller objects of the solar system. As stated in my prior books, if we only depended on solar light, we would not have any weather at all. Earth would have no winds, clouds, rain, cyclonic storms, jet streams, or lightning. All of our weather and that of the other planets is dominated by electric currents from the solar capacitor.

XIII.
HOW DO THE PLANETS FORM?

THIS IS THE MOST IMPORTANT CHAPTER IN THIS BOOK.

If you come away with nothing else, the information presented here is paramount. The completely failed theme of standard science on how the solar system and planets formed is so pervasive that, even though you do not realize it, it has become part of the worldwide human psyche. The biggest takeaway that could not possibly be true, is that somehow the universe and our solar system formed long ago and nothing has happened since (except for a few comet or asteroid strikes).

Every ten years the same incorrect theme arises that somehow a comet hit Earth and caused all the damage that we know happened in the past 10,000 years of man's existence. This is controlled opposition at its best. The news media picks up on this and promotes it as the perpetrators of this rehashed gain international fame AND FORTUNE.

One such wave was the book and public display of Clube and Napier in 1979. A barrage of comets allegedly hit Earth, causing all kinds of chaos. Again, it was allowed to have traction in the controlled news media as controlled opposition. "An asteroid or comet hit it" remember is an acceptable explanation especially, for the controlled opposition. It is allowed.

Currently there are a number of authors airing the exact same theme, but the people giving it traction have no idea that this is not some kind of new idea. It has been thoroughly exhausted many times in the past. Controlled opposition gets funded many times to very high levels.

Where does this funding come from? Usually, it is in the form of front private agencies, endowments, or foundations. If you follow these to their root origins, they all lead back to the CIA and laundered drug money or other secret operations that gain a lot of money behind the scenes. Once the money backs these unsuspecting pawns (who think they are getting money because of their fabulous research), they are off to the races. World tours, radio programs, Netflix or History Channel specials, book signings, and the list goes on. Money produces more money and the end result is that the real story never arrives at the table of the unsuspecting public who does not realize that this is just the same old story rehashed on a new palate for their entertainment.

Eventually the NASA and standard science white hats will come riding in to save the day, as the "alternative" explanation for the evolution of man has been squelched many times before. All of this can take years to play out. The biggest issue is that, in the ensuing milieu, the standard story of evolution of the planet Earth and man are never breached. The colliding comet or asteroid theme just indicates a slight hiccup in the standard story. The real story of how the planets form is never breached.

By now you are asking, what is the real story of how the planets form? The real story is that the solar system is a dynamic place and is forming new planets all the time. Comet Hale Bopp is an example of a planet forming from a comet and NASA successfully was able to hide this from the public through well orchestrated disinformation schemes. They even created a crazy lady who claimed to talk to aliens to represent the controlled opposition on the theme of "Planet X." New entries that actually become planets are a rarity because they take so long to form and spend most of their time outside of the visible solar system.

There are now cataloged thousands of newly forming planets in a region known as the Kuiper Belt. This is a doughnut shaped region beyond the planet Pluto in which literally thousands of planets have been discovered. Some are said to be Jupiter size but with the majority being smaller like Pluto. At the beginning of this book is a photo of a nearby star with its nebular ion cloud that is the equivalent of our Kuiper Belt. The lesson you learned in grade school that taught there were nine planets was totally incorrect. Rather than admit this massive flaw to the public, the astronomical community instead opted to cancel Pluto as being a planet.

Oddly, this was discovered during the heat of the Planet X massive disinformation program whose entire process was to squelch the idea of there being any other planets besides the original nine. By demoting Pluto from being a planet, astronomers relieved themselves of the embarrassment of going from nine to thousands of planets overnight. No one in the science media bothered to ask the obvious intelligent question .... "Where did these thousands of new planets come from?" Certainly, the best laid theories of the heralded peer reviewed journals never conceived of such a situation. The answer is that I, alone, had predicted this in my original work in 1979, which was trounced upon as I was publishing my papers as a young faculty member at Cornell University.

My 300 page book describing this entire process was entitled "**Origin of the Planets – Comet Capture Processes in the Formation of Solar Systems**" (1979), which later became the basis for what is known as the 3 part comet paper (1981). This is reprinted in the appendix of my "**Planet X Comets and Earth Changes"** book (published for general public 1999).

My paper trail is clear on this topic and is spot on. I called the "Kuiper Belt" region by its correct name "the Nebular Ion Cloud" and predicted that it would be a region rich in planet formation using my Plasma Discharge Comet Model. It also predicted that comets would exhibit x-rays to the sunward side (discovered by NASA in 1996 and rated as one of the top ten discoveries by *Discovery Magazine*). NASA hid this and claimed to have found an after-the-fact effect called "ion exchange," which was recently debunked by scientists from Europe who re-examined the NASA data and confirmed that the x-rays discovered were, as I had predicted, (Bremsstrahlung or braking x-rays).

Now you will understand why every few years a new batch of "an asteroid hit it" impostors are allowed to emerge as controlled opposition and promoted. This is underway today under the names of people like [name removed] (a non-scientist who is taking on the roll), and the list goes on. The people giving him air time do not know that all he is doing is rehashing the same old "an asteroid hit it" scenario that has been promoted and ultimately died on the vine dozens of times in the past. The "Planet X is coming – the sky is falling – chicken little" misinformation campaign is alive and well on the YouTube channels with wild predictions occurring almost daily. The same people recycle the story over and over again and when each prediction fails, they wait a few months and then go at it again; the same predictions being recycled over and over again.

The CIA has even gone so far as to edit short clips from my early 2000s radio shows and created a YouTube video complete with weird space music making it sound like I am predicting the imminent approach of "Planet X." This is recycled every now and then and I have to write to YouTube to take it down, which they usually do not comply. They are as much a part of the problem as the impostors that post on YouTube. For this reason, I have a posting on my web page that specifically states that I do not and never will post on YouTube. If you want to learn how to fix your bike or learn about a place to travel to, go to YouTube.

If you want science DO NOT DEPEND ON YOUTUBE NOR WIKIPEDIA NOR SOCIAL MEDIA. The American public has been so misinformed that attempting to right their direction is nearly impossible.

The CIA has even created groups that claim the same "discoveries" as my previous well established work, with my paper trail going back to its origins in the 1970s. These Johnny-come-lately types are promoted and well funded, but are nothing more than exactly what I have been talking about above. They are not scientists nor do they have any science training of any kind. These are the controlled opposition claiming "electric universe" or other themes that I developed. They have handlers who control their message and clog the internet with garbage.

Getting back to the subject at hand, which is "how do planets form?" All the planets, except for the sun's twin star Jupiter, were captured one at a time and settled into their respective orbits by the action of what I call "Tail Drag." This topic is covered in my other books that are required reading. The unfortunate situation today is that everyone wants the tik tok fifteen second answer. This topic does not lend itself to fifteen second nor even fifteen day sound bites.

As these new planets move in amongst the existing planets, they can cause tremendous havoc by close passages using what I call "Action at a Distance." There is a vast array of effects that I have covered extensively in my prior books. It depends on how close the newly forming planets (in the form of large comets) come to the Earth, Mars, or other planets as to the damage that is affected. These include gravitational, electrical, magnetic, pollution events (the great flood and other events where material rained in from outside the planet). The main issues here are that these comets do not hit Earth but only pass nearby. The Action at a Distance can be in the form of pole shifts, ice ages, rising sea levels, influx of hydrocarbons (oil as the ancients call it "naphtha"), electrical effects, and much more.

The main points to remember are that these events happen at random (there is no regular pattern) and many times have little or no effect on Earth. In the collective recorded history of the development of man and its many societies, there have been many disasters caused by passing planet-sized comets. This is why the powers in the world want to continue supporting scientists who claim that comets are harmless little snowballs. This is allowed science. "An asteroid hit it" is allowed science. The real story is not allowed because it threatens their power structure.

The reality is that we have very little history of these events and since no one is allowed to study them (other than the occasional "an asteroid hit it" researcher), it never sees the light of day. The upcoming chapters will delve into such topics as the true age of the Earth and the fact that Earth is gaining both water and solid material all the time from outer space. This causes a layering of material such that the surface which was here ten million years ago is primarily buried under huge amounts of acquired mass. Only occasional outcroppings are observed.

If this were not true we would be finding dinosaur and mastodon bones in every back yard (since they covered the entire earth) but we do not. Only in rare locations do we find such remnants. The idea that a single comet or asteroid hit Earth, causing all of the damage we see, is beyond preposterous. There are no major craters that support this idea and the ones that are discovered do not show mass extinctions even locally, let alone across the entire globe.

XIV.
WHAT IS THE TRUE AGE OF THE EARTH?

Standard science (or should I say the standard science religion?) recites in a nursery rhyme manner that the solar system and all its members formed 4.5 billion years ago. How exactly did they come to this astonishing conclusion?

As it turns out they picked up a rock on Earth and allegedly a few from the surface of the moon and via "dating" methods that are in themselves suspect and based on assumptions that the universe is subject to slow evolution, they came up with this number. It also is based on the assumption that everything formed at one time. This is an assumption that has no foundation in any form of experimental fact finding.

It is circular reasoning at its best. An assumption is proven by taking data and interpreting it based on the assumption. Once the date was set, then all the standard tier 2 scientists have had to do is to repeat it until it became science fact. As long as everyone in the choir is repeating the same verse, the choir gets funding, publications, and scientific vacations. Not a bad deal if you can get it.

When I attend international meetings such as the American Geophysical Union meetings, there are over 20,000 scientists from around the world. They all parrot the same dribble and none of them have a clue that they are the smoke screen of tier 2 science.

They have no idea that there is a complete layer of real science above them.

Central to the development of man is the concept that the Earth is an ever changing place and that its history is marked by violent events that have reshaped it in all of its myriad aspects. The surface, mountains, atmosphere, magnetic field, oceans, salinity of the oceans ... the list is nearly infinite. Colliding asteroids or comets do not explain any of this. We do not find many impact craters and even the largest show that the life and fauna lived through it with no great changes. Life continued unabated on top of the craters just as before (remnants found under the debris of the craters).

The real answer is that we simply do not know and cannot know the real age of the planet and solar system. The Earth has been building up layers upon layers. It has had many great floods, which includes not only great influxes of water but also other pollutants such as hydrocarbons (oil) and thousands of other chemicals that are abundant in the tails of comets. The Earth passes through the tails of comets, which is just one of the many devastating aspects of comet activity.

Since the sun's fusion is not in its core, the "calculations" of burning hydrogen in the core are meaningless. The sun is continually replenishing its supply of hydrogen from the galactic environment and therefore there is no way of knowing how long this process has been going on. Scientists, archaeologists, anthropologists, geologists, etc. all love to put dates on things when they have no basis for doing so. It makes them look smart. It fares better than saying "we have no idea."

We can state that the Earth is very old and has had many changes in its history. The changes we are most familiar with are what I call the Venus Event (or Moses Event where the Israelites were led out of Egypt) or otherwise known as the Velikovsky event. Then there is the downfall of Atlantis, which may be a completely separate event.

Another completely separate event is the Noah's Great worldwide flood, which was stated to be forty days and forty nights of torrential rain. Note that the Venus event and the Noah's flood event have completely different effects although both were caused by passing comets. The Venus event was the result of a large passing comet that was relatively near to Earth, whereas in the Noah's flood the Earth simply passed through the comet tail at a great distance from the nucleus. See my other works dealing in detail with these events, including my 2 CD set with 18 hours of MP3 lectures entitled "**The Physics of Ancient Celestial Disasters.**"

The various Ice Ages were also caused by passing comets giving an influx of water and cooling of Earth. We, in fact, now are coming out of one of those ice ages. The Earth is experiencing global warming and the worldwide glaciers are in the final throws of receding and melting into nothingness.

There is the reality of the mass extinction of the mastodons and woolly mammoths and a large array of large animals on Earth. Man was well established on Earth at the time of this Earth-wide disaster and it is possible that this event corresponds with the Venus Event. One of the possible mistakes made by Velikovsky is that in the follow on works to Worlds in Collision, Ages in Chaos and Earth in Upheaval, he attributed all of what is seen as on Earth as catastrophic to a single event (the Venus event). This is the same mistake being made by other subsequent people making claims of worldwide catastrophes. He also put dates on these and, in my opinion, was the most prolific investigator regarding dates.

I personally do not put dates on anything. But we can determine a sequence of events and also determine that there have been many large Earth populations on planet Earth. For example, the writings are very clear that the pre-flood population of Earth was enough to have ravaged its resources to the point where the populations did not have enough to support the population. We also know that the Atlantean population was worldwide and that they had colonies, including in the region of what is now Egypt.

We have a very good estimate of the population of the Mideast and the advancements of the Egyptian pharaohs; however, there is growing evidence that they did not create the marvelous art work and architecture that is generally associated with the various reigning periods of the pharaohs.

Here again Egyptology is an area where dating has been placed on time frames and a rash of contradictions present themselves. Since it is apparent that the current state of Egyptology is incorrect in its dating and interpretation of the construction of all we see there, that means that there was an extremely advanced race that preceded them of which we literally know nothing.

Partial to all of this is not only the age of the Earth, but the dating of various epochs in the Earth's history that are separated by major physical events. There have been many such events. At present there are more questions than answers, but making up answers to look intelligent will not solve the real issues of dating man's past.

XV.
VIOLENCE IN THE UNIVERSE

The universe is filled with long lapses of serenity and peaceful calm. Occasionally, violent events occur that make radical changes in the landscape. Man's past has been littered with many such events. Standard science is bent on hiding them for their masters' protection. Of course, they have no idea that they are doing this. The average university professor has not gotten his nose out of a textbook long enough to feed his dog, let alone understand his roll of being the stooge for trivializing man's history. I call these professors "textbook repeaters" or "textbook geniuses."

The idea that the universe is grand and only an occasional little "an asteroid hit it" scenario in the distant past (65 million years ago) has scarred our planet is their mantra. The reality is that in just the past 10,000 years man's tranquil existence has been shattered many times. In controlled opposition the "researchers" are allowed one instance of "an asteroid hit it." Watch the news and internet now you will see it along with the never ending barrage of "Planet X is coming tomorrow" nut cases. This is the disinformation world we live in. The idea is to flood the public with BS until they simply turn it off and go back to their menial jobs and pay their house mortgage and insurance bill and give control of their lives to the government who they believe will solve all of their problems.

When I first read the Kolbrin opening dialogue about the renovation of the human race, the hair stood up on the back of my neck. The Kolbrin is a series of ancient texts salvaged from the personal library of Joseph of Arimathea (a wealthy roman businessman). It is clear that this information was common place among the educated populous around the time of Christ. It describes how the great comets devastated planet Earth and the effects that are all characteristic of a large passing comet.

The reason one puts credibility in ancient books, such as the Kolbrin, the Old Testament, the Mayan Popul Vuh, Plato, or so many other writings is because the ancients describe vividly and in great detail the events that we now can interpret with science that they could not have known about. They talk about interplanetary electrical discharges, comets causing tsunamis (Quetzalcoatl carrying the ocean on his back), the seven plagues of Egypt (all explained by comet effects) and hundreds of other phenomenon that they simply could not have invented.

Add to this the fact that the various cultures from around the world all concur on the same details, and we have indisputable evidence. **Since modern astronomers, like Carl Sagan and his merry band of traditional scientists, could not match these writings with their outdated incorrect theories, they concluded that the ancients were misguided.**

My work is essential to put ancient writings into perspective, confirming that they knew exactly what they were saying and that the violent history they portrayed was in fact correct.

The universe can be a violent place, but this is by design. If it were not for these events that stress the species, the world we see today would not exist. This is part of the grand design.

XVI.
COSMIC and DARWINIAN EVOLUTION

THE FALSE BASIS OF MODERN SCIENCE

Slow cosmic and biological evolution are at the heart of the Sagan COSMOS series. Darwin's leap of faith of slow deliberate changes to evolve new species follows the same map. The real situation is much different. The shark "species" has been the same since the first shark fossil has been discovered. No one really can tell what other species it may have evolved from. Of all the species on the planet, only variations within the species have been found. The magical and elusive "missing link" has never been found, not only for humans but for literally all "species."

The strange part of the phrase "THE missing link" implies that just one would be needed to fill the gap. "THE" missing link? There should be thousands of them and they should be all over the place. There may be a small gap here or there, but as far back as we can go in history we find man in his current form.

In my book **The Diamond Principle** I pose the following question. If you find a diamond on the far side of the universe, will it have the same properties as a diamond here on Earth? Every individual diamond is different with internal fissures and imperfections, but the process that creates the diamond will be the same.

If you find DNA on the far side of the universe,

will it be the same as you find here on Earth? And if you find biped "intelligent" beings on the far side of the universe, will they resemble man here on Earth? Of course, you then ask if you might find a red oak tree on the far side of the universe? Of course, this would not be expected, however there will be expected trees and plants using photosynthesis.

This begs the question, "did man originate on this planet?" The idea that the universe began in a Big Bang conveniently does away with all kinds of other concepts regarding man. It denies our existence as a creation of a creator, it denies our spirit and creativity and individuality. It is the perfect communist atheist manifesto. In one fell swing it destroys the entire purpose of man.

Is the universe finite? Every time we build a new telescope, we see more galaxies filling what we thought preciously was void. None of the theories taught today could possibly be true.

There is an astronomical lemma that incorrectly states that if the universe were infinite, then you would see stars covering the entire sky at night. This is posed by astronomers who are not very astute mathematicians. Mathematics is full of infinite sequence equations that do not add to infinity. Likewise, neither your eyes nor the most powerful telescope is capable of seeing beyond a certain point. The real issue is that man's finite mind cannot envision the infinite in terms of time or space. So man creates a beginning and an end in time and space to the universe. Cosmic and biological evolution are psychological crutches for man's finite mind.

XVII.
HOW MANY AGES OF EARTH HAVE THERE BEEN?

The best answer is that we simply have no way of knowing. Hundreds and even thousands or maybe more would be a good assumption, given that we know of at least five major Earth wide devastating events in just the past 10,000 years. More importantly is the fact that we know enough now to formulate the question.

We know that the planet Mars had a different orbit that was altered when the large comet Venus passed by. Mars was a blue planet. In the world of standard science, with the planets forming all at one time 4.5 billion years ago, no one had to bother with such realities. The ancients record how Mars lost its atmosphere and oceans in a single night. I discuss this at length in my prior books. Has Earth also lost its atmosphere or had it severely altered by a large passing comet? The answer is definitely yes. Mass extinctions are generally accompanied by great changes in Earth's atmospheric makeup.

Standard science sweeps such inconvenient facts under the rug, assuming that Earth somehow did all of this by itself. The real history of man and planet Earth cannot be discussed without the major events that were instrumental in shaping his past and destiny. I once was in the house of an honors history class teacher of a large metropolitan high school. I picked up the history book for honors seniors and flipped to the beginning. It recounted the standard story of man evolving in the valley between the Tigris and Euphrates rivers, who then emanated to populate the Earth and evolve into all the races of man we see today within the past 7000 years. This story is so outrageous it could not possibly be true nor explain what we see today.

I asked the honors teacher where the part was about the great flood and the teacher replied, "Oh we don't teach any of those fairy tales any more." Another incident: I was in a church service, and upon leaving asked the pastor about the Old Testament and the great flood. He told me that they no longer teach that since the bible was full of fanciful stories that could not be true.

The reality is that the human race has completely lost track of its origins and therefore is doomed to relive these same events. Man is doomed to be driven asunder again, and when it happens no one will have the slightest clue of what is really going on.

There are today many groups concentrating on what is called the "Younger Dryas" impact theory. It is their belief that a single comet crashing into Earth around 12,000 years ago was responsible for all kinds of changes to the development of man, including the extinction of species from the Holocene period (mastodons, woolly mammoths, saber tooth tigers, giant sloth, and more). The only issue is that it does not explain dozens of things, like the tropical mastodons flash frozen in the tundra of Siberia standing on all four legs with fresh undigested tropical flowers. The comet impact cannot explain a pole shift. It also requires that one identify the impact site and examine the fauna and life systems both below and above the impact crater. To date they have not even identified the alleged impact crater, so there is a whole lot of trumpeting and fanfare before there is anything to cheer about.

XVIII.
HOW MANY TIMES HAS THE EARTH BEEN DEVASTATED?

The answer is a relative answer. What do you consider to be "devastated?" Will a partial mass extinction even do? Do you need only a minor influx of poisonous gases to qualify as a devastation event? Or will a simple influx of energy from a planetary comet electrical alignment, which causes sixty hurricanes on Earth at one time, enough for your definition?

In 1999 there was a small comet called Comet Lee that was coming into the solar system, which I discovered was going to be involved in a direct electrical alignment with planet Earth and a number of other planets. This meant that on a certain day ... September 6, 1999, there was a high probability of a large discharge between the planets, comet, and Earth. To bolster this event was what I call the New Moon Passing (two days after full moon when the solar wind comes crashing in to our protective plasma magnetic field and acts as a trigger to initiate ionosphere-to-ground discharges and to ignite cyclonic storms).

In May of that year, I made the prediction that this date would have a lot of hurricanes. As the date approached my detractors started to hoot and howl on the internet that "McCanney has been proven wrong as there was not enough time for major storm systems to start." The Earth wide weather was very calm and no hurricane activity, in site up to and including the day prior to that date. At about midnight on the designated day, over sixty full blown hurricanes ignited, including five in the Atlantic Ocean with one of these threatening the state of Florida. As I like to say regarding the government paid detractors ... the silence was deafening. I had successfully predicted something that no one could have imagined.

Earth is under constant stress from the sun, comets, planetary electrical alignments and effects, such as tidal waves in the oceans and even the Earth's surface. Jupiter, for example, has a thirteen month cycle in which we align electrically, and Earth experiences what I have termed The Jupiter Winds. These are straight line winds that blanket the Earth for about a week to ten days. The US Military satellites first measured the electron beam connecting Earth and Jupiter in the 1990s, and again my prediction was 100% accurate dealing with the discharges crossing the solar system.

Another form of potential disaster by the hands of comets is that of viruses forming in the tails of comets. I had been talking about the cycle of flu viruses that were first identified by the ancient Chinese circa 1250 AD. They realized that during the inferior conjunction of the planet Venus there would be a large influx of influenza. This was well accepted and is noted in the modern medical peer reviewed journals. This would also imply that during comet encounters, Earth would likewise be subjected to an influx of viruses.

Besides this, it has been noted that sun spot

activity increases also at the inferior conjunction of Venus with Earth. But what other effects might befall Earth during a near passage of a comet encounter?

In my book **The Diamond Principle** and also in my prior writings (Calculate Primes) I noted that diamonds form in the sunward side of comet nuclei by the following process. In the laboratory micro diamonds were formed by vaporizing carbon to atomic form and, by adding a current of electrons, small diamonds formed. These same conditions exist to the sunward side of comet nuclei. In the study groups that are researching the Younger Dryas event, they are seeing micro diamonds peppering the ground around sites. They, however, claim that this is due to a comet colliding with Earth.

The real reason is that Earth was barraged with diamonds that streamed into Earth as the comet passed by Earth (not a direct impact). As noted previously, the one impact controlled opposition position is allowed to prevent the real situation from coming to light. This keeps it at one event long ago (12,000 years) … out of site out of mind. The reality is that every time Earth is besieged by a passing comet, micro-diamonds spraying Earth would be one of the resulting telltale signs. This has happened many thousands of times and to the unsuspecting person this may all appear to be a single incident.

XIX.
HAVE THERE BEEN ANY RECENT NEAR MISSES TO PLANET EARTH?

I have repeatedly stated in my lectures and in my books that there have been many incidents of Earth wide disasters from comets that passed nearby and did not directly impact Earth. These rare planet sized objects are the core nuclei of large comets (most comets are very small). One has to make the transition from thinking that comets are comets and planets are planets. The Earth and all the planets continue to this day to discharge the solar capacitor and have the same characteristics as comets in elliptical orbits, albeit much less active. These planets formed from active comets that raged through the solar system during their formative periods.

There are thousands of other objects that form comets as they come into the solar system and discharge the solar capacitor. By astronomical definition, we call all of these "Planet X." This is the astronomical term given to planets before they are officially given a name. These must be recognized as the MANY Planet X Objects.

There is one particular Planet X object that was searched for and discovered by Robert Harrington of the Naval Observatory throughout the 1980s, resulting in the discovery of what later became the comet we call Hale Bopp. The Planet X we are dealing with here became the comet Hale Bopp, which has its own history. It passed through the solar system with little effect on planet Earth. The entire history of this famous modern comet is covered in my last book "**Planet X Update 2020 – The Real History of Planet X**." It is now leaving the solar system and will be back again on its Earth crossing orbit in about 2600 years.

The reality is that when Harrington first discovered this planet X object (literally labeled as Planet X under contract from NASA) it was on a near direct collision course with Earth. Because of the tail drag it lost energy and Earth moved ahead three months in our orbit. As a result, it missed us. If we would have been a mere three months earlier in our orbit, Earth would have experienced a near passage and undoubtedly a mass extinction event. Hale Bopp would have passed a mere million miles away (about four lunar distances).

Will Hale Bopp affect Earth on the next passage? The answer is difficult to estimate or calculate because it all depends on the condition of the solar system at that point in time, the orbital location of Earth, and the unpredictable effects of tail drag that perturbs the comet orbit. Will there be a large solar capacitor to discharge that will affect the level of tail material accumulation, which will affect its orbit? One thing is certain that it will again follow an Earth crossing orbit and is a comet of major proportions.

Does the intelligence community have the right to withhold this information from the public?

Representing the World Bank that considers the people to be resources to be used as needed or discarded as needed, we can see that the answer is clear. The main reason is control of the people … if the public realized there was a real threat, then they might stop paying attention to the evening news and stock market and housing crashes and insurance payments and on and on and on. The Public may wake up and shed its leaders like a moldy old suite. They would also begin pursuing outer space in earnest, to save the human population and our genetic structure for posterity.

When the minor planets in the Kuiper region started to be discovered, it created somewhat of a stir in the astronomical community. The problem again was that the data contradicted the long heralded idea that the nine planets from Mercury to Pluto all formed at the same time 4.5 billion years ago, and nothing else existed until you might happen to the next star (except for the alleged "Oort Cloud" where allegedly icy comets had formed and lived). The problem only increased with time as over 40 new planets had been discovered. Eventually the list grew to over 1,000 new planets with only a small portion of the sky having been examined.

Instead of admitting the failure of the "Nebular Collapse Model" of solar system formation, astronomers decided to demote Pluto from being a planet by a vote. It would be easier to convince the unscientific public that Pluto should never have been classified as a planet in the first place, than to admit that now there were 1,009 planets instead of the textbook nine planets. The ugly topic of Planet X enters here with the oddity that if Pluto (Planet number 9) were removed from the standard list, then instead of locating Planet X in case a new large planet was discovered in the Kuiper belt, they would now be able to name the new planet as Planet Nine and avoid the problematic term Planet X (Planet Ten).

We know that NASA has discovered hundreds if not thousands of planets associated with our solar system and stopped reporting them as the discovery rate turned exponential. One statement said that there are more planets outside of our visible range than planets associated with our solar system and that this was the general case with all stars … not the exception. There are not only little planets but some of Jupiter size planets.

So much for the solar system forming all at one time into nine nice neat planets 4.5 billion years ago. This absurd fairytale still fills the textbooks today from grade school to highest level university class material. Astronomers recite it like an old Mother Goose nursery rhyme. Data is reported that completely contradicts what I call the "Axioms of Astronomy." The Big Bang, Icy comets, electrically neutral solar system and 4.5 billion year old nine planets … ALL FALSE … but don't tell that to the fake science news media or the textbook companies.

There are literally hundreds of Planet X objects associated with our solar system but the caveat is that very few, if any of these, are Earth threatening in the least … at least not on a time frame that any of us will have to worry about. From statistical calculations based on known planet numbers, it is certain that there are many planets that will come into our solar system over the course of millions of years. We have no record of these other than knowing that Earth has been devastated many times in the past and it will happen again in the future. This does raise the question of time. Given enough time, some of these will present major issues for our peaceful planet.

XX. FINDING OIL and WATER MANY MILES UNDERGROUND - THE TRUE CONDITION OF THE MOON, MARS AND OTHER PLANETS

Earth is a layered planet. On a daily basis we have material coming in from outer space. In the 1980s a scientist named Lou Frank, one of the original founding scientists of the van Allen Belt team, placed a UV detector on a satellite and took data showing that Earth's upper atmosphere was full of small patches of water. At first, they imagined that this was due to small micro comets but that did not explain what they were seeing. Instead of small micro regions there were wide areas with influxes of water.

The ancients tell us of great influxes of water and hydrocarbons (petroleum) in addition to influxes of solid material up to bolder sizes. Since Earth is acting like a comet, all the time discharging the solar capacitor, it is continually bringing in everything in our vicinity of outer space. This is a direct result of my Plasma Discharge Comet Model. The Moon and Mars are the same.

When oil was first used by man, they found it in pools on the surface of the ground and learned how to distill and refine it to some degree. They used it for warfare and for fires. When the surface pools ran dry, they began to dig down and found more. The rest is history, as today we are drilling down to find oil up to 12 km below the surface. This has led to the belief that oil is created at these depths (called abiotic oil). This is a bizarre concept since by its name it claims to replace what is of itself a farce (fossil fuels). The idea that oil comes from dinosaur bones and fern forests is patently false and was promulgated to make it sound like a scarce commodity.

The reality is that neither of these are true sources of oil. Oil forms in the tails of comets. Earth has repeatedly passed through the tails of large comets. The influx of oil and water have always been part of the pollution event. As Earth accumulates more and more matter, these deposits of water and oil simply become buried.

XXI.
EARTH CHANGES AFTER A LARGE PASSING COMET - POLE SHIFTS - MOUNTAIN BUILDING - ICE AGES AND GLACIERS - ATMOSPHERIC CHANGES

My book "**Principia Meteorologia – The Physics of Sun Earth Weather**" goes into detail on this topic and not only describes the effects on Earth for large passing comets, but the other planets as well. Also, my book "**Surviving Planet X Passage**" has a detailed "before, during, and after" explanation along with a **Frequently Asked Questions** section.

My book "**Atlantis to Tesla – The Kolbrin Connection**" as well as the original book "**Planet X Comets and Earth Changes**" all go into detail on the effects of comets. The last books mentioned also have appendices that reference my work and paper trail going back to the 1970s, including some papers published in peer review astrophysics journals.

The main point is that the effects are caused by comets that do not crash into Earth, but from passing comets. I have a series of lectures that include the various effects as one might change encounter parameters, such as relative velocity, electromagnetic effects, gravitational effects such as tidal waves (tsunamis), and pole shifts as well as mountain building.

The ancients are very graphic in their description of the passage events. For example, the Maya describe Quetzalcoatl, the plumed serpent of the night sky, as carrying the oceans on his back and causing great devastation. The same story comes down to us from all parts of the world.

The Kolbrin equally displays the fighting sky monsters doing battle with planet Earth. The ancients describe all of this and more. The error that is made by literally every researcher is that they claim there was only one event. As noted, they are only allowed to claim one event and it has to fall into the realm of "an asteroid or comet hit it." Brace yourself as the media will promote this beyond your wildest dreams and drown out any other voices.

XXII.

WHAT THE ANCIENTS DESCRIBE- QUETZALCOATL - THE PHOENIX- - CHINESE DRAGON - GREEN RIVER, UTAH PETROGLYPHS - THE VENUS EVENT - ATLANTIS- NOAH'S FLOOD - MASS EXTINCTIONS - RAIN OF FIRE FROM ABOVE - THE END DAY PROPHESIES

The following photo is from the cover of my book "**Planet X Comets and Earth Changes**." It depicts a snake like object that appears to be ready to attack a planet. More than that, there are two large comets. Both exhibit large spiraling sunward spikes and the one on the right shows the nucleus. Given that the average size of a large cometary atmosphere, this could be a planet the size of Jupiter OR LARGER. The artist left us this important data.

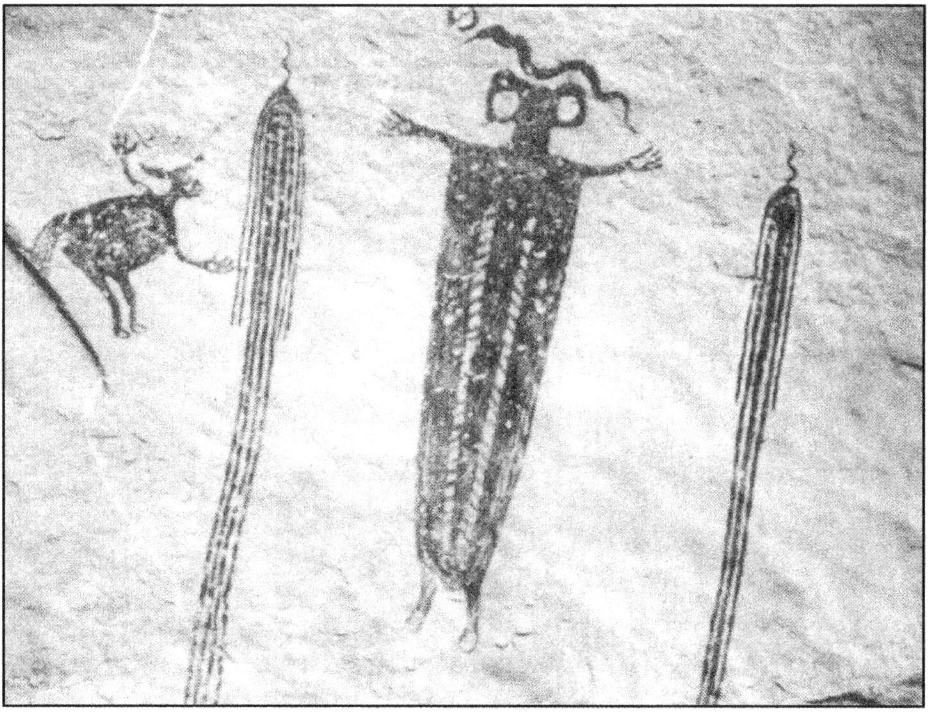

The images on the following pages constitute a small sampling of imagery passed down to us from the ancients from around the globe.

QUETZALCOATL – the plumed serpent god of the night sky whose heart became the planet Venus (Central deity of the Maya and Inca also adopted by the Aztecs)

THE PHOENIX American southwest image of the god that destroyed Earth from the heavens and it is said that the Earth regrew from the ashes left by its passage.

CHINESE DRAGON – celebrated as traditionally symbolize potent and auspicious powers, particularly control over water, rainfall, typhoons, and floods. The dragon is also a symbol of power and Earth regeneration

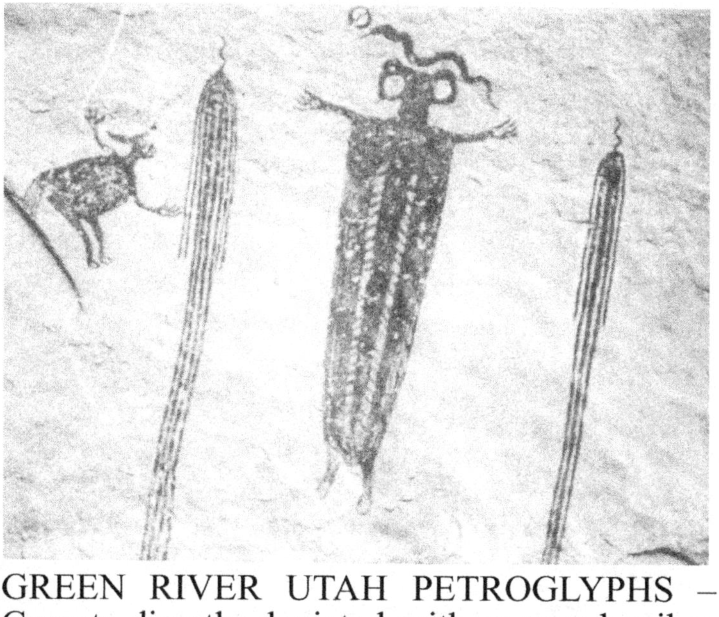

GREEN RIVER UTAH PETROGLYPHS – Comets directly depicted with sunward spikes and large planet sized nucleus.

THE VENUS EVENT - Immanuel Velikovsky myths handed down from all over the ancient world and the Kolbrin imagery and ancient Mayan Art

ATLANTIS – the 5 ringed city is flooded

NOAH'S FLOOD – 40 days nights torrential rain

MASS EXTINCTIONS – the mastodons flash frozen in the Siberian tundra standing on all four legs

RAIN OF FIRE FROM ABOVE –
Sodom and Gomorrah

THE END DAY PROPHESIES – the stars will fall from the heavens (meteor streams from comets) the moon will turn to blood (volcanic eruptions on the moon) the firmament will be shaken (gravity waves from passing comet cause Earthquakes)

These are just a small number of real events from the past few thousand years on planet Earth that have changed the course of man's development and the planet we call Earth. The end day prophesies are based on past events that will repeat in the future.

Note that all of the above historical events happened within the past 10,000 years. These all occurred long AFTER the alleged "Younger Dryas Period" comet collision theory. That idea puts all of these in one pile 12,800 years ago. Out of sight and out of mind. Do you see the construction of the misinformation placing it long ago with a theory that attempts to give credence to the failed dirty snowball comet model of NASA and standard astronomy? This has all the earmarks of a well orchestrated disinformation program that uses unwitting players as the messenger.

XXIII.
HOW DID HUMANS RESPOND TO MASS DEVASTATION?

We have direct data as to the response of humans to events that are so devastating that it shakes them to the very core. The Kolbrin first chapters are graphic if you understand what they are saying. It recounts the rebirth of man during the terrible days of passage of a large comet. The people who survived emerged from caves in the mountains and were deaf. The tremendous noise and trumpeting destroyed their ear drums. They were barely able to sustain and could not relate what had happened to their children other than in the most rudimentary ways.

Myths and legends descended and that is why we have very little to go on today. But great scholars have routed out this small amount of information. The human race nearly went extinct on many occasions over the past 10,000 years. The ancient Greeks formed new governments based on logic and natural law. They developed language, art, government, laws, science, mathematics, education, sports, and ethical standards all of which have influenced our cultures to this day.

The human spirit is resilient but has some fatal flaws. We tend to forget history that should be guiding our future. The ancients certainly did not have a record of discussing outer space (they did not know about outer space as we know it today).

Immanuel Velikovsky, a clinical psychologist by trade, coined the term "collective amnesia" to assign a clinical designation for the condition experienced by ancient cultures. Like a traumatized patient, the event was so horrific that man put it behind a psychological door.

Leaders additionally put the lid on the public knowing about such things because it would instantly erode their hard earned power structure. The only solution would be for man to become space faring and to save the DNA from extinction. Today we rest at a crossroads, which leads us to the final two chapters of this book.

XXIV.

WILL MAN SURVIVE?

The rules of nature are quite simple, and in the case of survival as an intelligent species (one that is capable of altering and controlling its environment) the rule of ultimate survival is clear and very simple.

As a species … collectively … if you do not use the time and resources allotted to you to gain access to outer space and gain the knowledge to survive … you will be doomed to extinction.

This clearly leads to some secondary basic rules of nature regarding the use of time and resources.

Petty wars and the use of resources to fight these wars can have only one possible result … and nature in its own time and manner will repay those species … collectively … who fail to control their leaders who promote these wars.

If there is a smaller group that attempts to use the resources of the many to save their own few and selective hides … these especially will be doomed to die a most lonesome and hideous demise.

The successful species that eventually will populate the stars will not allow those "unsuccessful" species to enter their hard-earned space and bring their petty wars and selfish ways into the cosmos. This is called the "rule of survival of the successful" ... and is universally understood and <u>commissioned to be executed</u> upon those who fail at the first three rules.

These rules are stated to be inherent and self-evident. As a species, we have failed. There are no less than 54 wars ongoing at the time of this writing. World leaders are out of control. As an observer of the Sun and comets, and a person with inside contacts to very high levels of the intelligence agencies, I will say that the writing has been on the wall for some time and there is little doubt that the leaders in these intelligence communities have been warned.

Let the reader decide ... based on the inherent rules of survival ... will man survive?

XXV.
THE REAL HISTORY OF MAN & EARTH

The final discussion of comets passing through the solar system and affecting the existing planets must center around the true nature of comets. For those who continue to imagine that they are fluffy little snowballs (especially in academic circles) they are only biding time. Their place in history will fade like so many other religions that have come and gone. The topic then moves to the topic of large comets. These are the type that have enough power gravitationally and electrically to cause major destruction when passing by existing planets, including Earth and our Moon.

The small comets may have effects but direct collisions are extremely rare (not to say they do not happen) but the passing large comet scenario turns out to be much more common place. This leaves us with the discussion of objects that form large comets (planetary sizes). We therefore revert to the topic of Planet X. This is a term that has been ravaged by internet cranks and impostors, many of whom have been directly supported and promoted by disinformation coming from the intelligence agencies. Planet X is a very old astronomical term that simply means any planet sized object that has not been given an official name. It is used in the astronomical community and is on NASA and military contracts and documents and used in this sense, not the internet nut case sense.

There is no topic that has received more bad press than the topic of Planet X. There is a reason for this and it took me a long time to figure it out. First of all, I have been dealing with this topic for all my many decades career, although the term Planet X, as it is used today in the modern press, is somewhat askew of the real astronomical term.

It is odd that the industry of astronomy would sacrifice a real astronomical term and allow it to be changed to provide a vehicle for misinformation regarding a topic for which they should be in the forefront ... objects from beyond our solar system. But modern astronomy had to take a stance back in the 1950s when it first confronted Immanuel Velikovsky and stated more than emphatically that there was nothing to worry about in our part of the solar system regarding items from beyond.

They denied that comets could affect Earth, because they stated comets were small little ice balls of fluffy white snow and the public certainly could conjure up the idea that ... yes ... puffy little snowballs could not hurt a flea, let alone Earth. Additionally, the man who rose up and finally, in 1974, confronted Velikovsky (Scientists Confront Velikovsky at the 1974 AAAS meeting) was a popular astronomer of the David Duncan Chair from Cornell University, named Carl Sagan. Just his stating something, with his cocked back overly confident arrogant manner with all of astronomy and NASA behind him, was enough to carry the weight of near God himself. He stated on his TV series COSMOS and in his related books and writings that the probability of an asteroid hitting Earth was so minuscule (remember billions and billions?) that it was a laughing matter to consider it.

The astronomical logic for this statement was that the then current belief in the astronomy community was that the bombardment of Earth and our moon with rocks and debris from space took place early in the formation of the solar system and nothing had really happened since. No sooner than the ink was dry on the pages when the astronomical community had to face a stern reality. The dinosaurs seemingly had an early demise due to an asteroid hitting Earth and a Nobel Laureate scoured the Earth for a consistent band of tell-tale iridium that only could have come from outer space and which was a deposit laid down on top of dinosaur bones.

The astronomical community quickly changed its story and rapidly embraced the concept of asteroids colliding with Earth. They, in fact, created a major NASA funding opportunity based on this with none other than Carl Sagan's favorite student Dr. David Morrison becoming the leader of the effort to study this topic. Major Hollywood movies even picked up the theme.

This all happened as I struggled to get my papers published in peer review astrophysics journals as a young Physics and subsequently Mathematics faculty member on the Cornell University campus. I was fighting on campus what I knew was totally incorrect … a more than David vs Goliath effort.

You must remember that during this time I had

no knowledge of the Velikovsky Affair as I had been out of the country for many years teaching Physics, Mathematics, and Computer science in a foreign university, so I was totally unaware of the hornets' nest into which I had stepped.

Another thing happened at the same time, which is one of the greatest scientific thefts in all the history of what is now known as the "Velikovsky Affair." An evolution biology specialist, named Dr. Stephen J. Gould, was one of the professors who ganged up to "soundly defeat Velikovsky" alongside of Carl Sagan at the 1974 AAAS meeting.

Carl Sagan, for his efforts at becoming the solidifying voice of the astronomical community and carrying the banner, got the benefit of becoming a Hollywood super star and the job hosting the thirteen-part TV mega series COSMOS (which became the most watched science show in history AND whose book is the most read book of all time second only to the Gutenberg Bible). Stephen J. Gould, on the other hand, got his benefit in that he robbed and renamed the concept developed by Velikovsky dealing with rapid evolutionary change of the species.

Gould could put the twist on the story that this was not due to impending disaster and violent change from outer space, as in the Velikovskian scenario, but stated that these rapid evolutionary changes were somehow spontaneous on Earth ... This was the formula needed to keep the house of cards in order and consistent with modern astronomical theory. Gould drew some simple charts and wrote some papers and was posthumously ordained prizes that went along with this "major discovery."

All of this was needed to pave the way for the public misinformation campaigns that would rage throughout the mid-1990s, as the large comet Hale-Bopp wound its way through the solar system, and for which my work was center stage in a battle against standard science and astronomy on a daily basis. My web page was receiving, at times, over a million hits per week and many of these were from government installations, including military and professional astronomical facilities. In spite of the denial pretended by these communities, they very much did not want the public aware of reality and the true nature of comets.

For some time, after certain events clearly illustrated that comets could not be dirty snowballs, the news media and even NASA went away and were quiet as a church mouse on the topic. But then all of a sudden, the lid came off and ever since all related articles coming from standard science have been 100% "dirty snowball friendly" and in most cases they cannot say enough times "little icy wanderer" or "fluffy interloper from beyond" etc., etc.

But something else happened. NASA built a barrier around itself in the news reporting sense. A man named Dan Golden came to be head of NASA, coming from the security end of the government alphabet soup agencies. His job was to put the lid on NASA. Workers and scientists were made to sign non-disclosure agreements and gag orders were placed on anyone with access to the least bit of data or real information. Only

"Principle Investigators" would have access to NASA data, which was now guarded like the Crown Jewels.

Divisions were made and created what I call the tier system of science. Tier I science and scientists were at the level where one will never see their names or work. NASA Tier I science basically joined with the military and were made essentially into one but, in true military style, all of the workers were compartmentalized. NASA created two news release headquarters regarding space exploration. They were located at JPL, the Jet Propulsion Laboratory in Pasadena, California, and the other was at the Goddard Space Flight Center in Greenbelt, Maryland.

Essentially, no one at NASA could get up and make a public statement without permission or going through the release agencies AND with very scripted statements only. No spontaneous statements were allowed nor interviews that were not screened by the central release agencies. A man named, Dr. Tony Phillips will generally have his name on all official releases. He is the final filter for news and information all things NASA. Standard people were listed as the news contacts. This was to keep the NASA story straight for public consumption. The Dirty Snowball Comet Model now regained its prior primary spot in the press and it has not left center stage since this time, no matter what the new data might say.

Another thing happened at this same time. Rather than have NASA make public statements, especially ones that could prove controversial or for which flack and defenses had to be mustered, NASA funded a series of what they call "outreach" personnel. This was a series of front people who were paid somehow through back door funding, but who appeared on the cover as "independent researchers."

These appear to be people just doing this out of the love of the "truth in science." These are people with no visible means of income, like on Phil Plait or Bill Nye the Science Guy, who pound the party line, battling the armies of disinformation stooges thrown out by the CIA. Remember that a good disinformation program creates both sides of the dialogue. These have been invaluable over the years to defend NASA but without NASA having to say a word.

This convenient arrangement includes workers, for example, at the Naval Observatory who comment in the press about comets always touting the standard accepted set of principles. It all looks so official. Their web pages look official BUT at the bottom you will always find in fine print that they are not associated with NASA. The Naval Observatory is NOT related to NASA that can conveniently stay hidden in the background and never has to answer any real questions.

This outer perimeter of scientists is what I call the Tier III scientists. They do not do any real research. Their sole purpose is to outreach to the public and they have a very limited scale of items that they talk about. They defend the Holy Grail of science topics and, as thus, I call them "textbook geniuses" or "textbook repeaters." They are the ones who get air time. Other members of the Tier III science are university teachers who repeat the textbooks that are presented for the Tier IV news media that blindly reprints exactly what is handed to them by the NASA standard news release agencies.

What I have learned from attending dozens of international scientific meetings is that the true history of astronomy and related fields is completely hidden from graduate students on their road to higher learning and getting their Piled Higher and Deeper PhD Degrees. They are green behind the ears, to say the least. One very telling incident occurred with the discovery of Sedna, the first planet discovered outside of the orbit of the then 9th planet Pluto, … discovered by an automated telescope search program and its "discoverer," one Michael Brown.

He was in a news conference with dozens of reporters. The belief was that Sedna was an icy body due to its distance from the sun, just like the beliefs about its inner neighbor Pluto (assumptions that turned out to be totally false since Pluto has an atmosphere and even weather systems of nitrogen compound forms of snow). Be that as it may, at the time the assumption was that Sedna would be an icy body covered with ice. The question was posed by a Reuters reporter as to what would happen if Sedna were to enter the inner solar system. To this Michael Brown replied that this would be the biggest COMET the world had ever seen.

In his youthful ignorance and exuberance to embellish the response, Michal Brown had just breached the topic that was taboo. This sent shivers up the spines of his doctoral overseers. NEWLY DISCOVERED PLANETS BECOMING COMETS – INNER SOLAR SYSTEM – WHAT IF IT APPROACHED EARTH

– PLANET X ... everything the entire astronomical community had worked decades to cover up had just been breached in a few short seconds by a young green behind the ears astronomer who had no idea what he was admitting.

The news media had to back pedal hard to smooth this over. Remember, this was a live interview and so no one could put the genie back in the bottle. Besides this, it was clear that the automated searches were turning up all kinds of planets. The astronomical community, along with its compliment science news media, had become adept at covering up reality. This was just one of hundreds of examples of making changes while keeping the "Axioms of Astronomy" in prominence.

Tier II scientists are the ones that everyone imagines are the top scientists at NASA and in our university systems, and with many in industry jobs that are government funded. There is a great expense to funding and maintaining the Tier II levels of science. These are the scientists who populate by the tens of thousands the international meetings (some of which I have attended and presented at) such as the American Geophysical Union meetings that occur bi-annually.

Last, but not least, is the enormous entourage of internet wannabes and impostors, many of whom are directly promoted and sponsored by moneys that come out of the black ops institutions that wish to control public thought on almost every topic, including the topic of Planet X.

But what does all this have to do with Planet

X? The answer is EVERYTHING!!! The reality is that there is an extensive, ever-pervasive, finely-tuned network of misinformation and what I call "dog and pony shows" that greets the unsuspecting new visitor to this subject. If you can wade through all the flack that has been generated for you then you will be one of the fortunate ones. There are more wannabe con men and small time little ego impostors setting in their Grandma's basement posting on the internet than you can imagine. The topic of Planet X has no lack of contributors.

The government, which desperately wants the public to remain very stupid regarding these topics, spends a great deal of time promoting the nut cases and their dribble. WHY? Because this allows them to control BOTH sides of the dialogue. They certainly do not want my name or work being promoted. They can blindly ignore my work and the publicity it would generate for me, while referring to the copycat web pages they support that twist and turn my work into a crazy topic run by truly the lunatic fringe cast of characters. This is an easy target to defend against by the Tier II and III NASA apologists, and in the public eye it is easy to criticize and defame the CIA sponsored crazy lady that claims to get her information by channeling with aliens on the topic of Planet X.

So, what is the real story of Planet X? First of all, Planet X is an astronomical term with a long and rich history by which many good well established renowned astronomers throughout history have seen what appears to be other planets in our solar system.

Until these objects are confirmed and given an official, internationally agreed upon astronomical name, they are politely called PLANET X. To begin with, Planet X is a historically correct astronomical term that goes back hundreds of years.

When I began the public release of my information in the late 1970s publishing in the peer reviewed astronomical journals, no one spoke of any Planet X objects, except possibly in passing, to identify the true history of objects that had been seen. There are hundreds of objects most of which never materialized as named objects.

But my work from its very nature and inception included the concept of Planet X, although it does not appear in my written work at least until the book "**Planet X Comets and Earth Changes**" hit the scene around the year 2000. My work centered on the concept that large comets could come into the solar system and become planets if the correct capture and other celestial conditions were correct. At best, this would be a rare event with many factors needed to coincide to make a new planet or moon. I used the title to gain attention to my work and the title was highly successful, although by this time the government sponsored misinformation programs had thoroughly made the topic of Planet X into something that only crazy people would talk about.

It was a two edged sword. I even considered calling the book PLANETS COMETS AND EARTH CHANGES, but I wanted to take the misinformation campaigns head on and change the public perception of this very important astronomical term and return it to its historical correct place. This is why I earlier mention that it seems strange that standard astronomy would allow one of its standard terms to be destroyed to make a point. You have to remember that destroying the concept of Planet X was not a decision of the science community ... it was a decision made much farther up the ladder, as I was to find out. The decision had to do with international control of the public.

To understand this, you have to understand who really controls the world and that it is done through deceit. An essential ingredient on this is that of which I call BAD SCIENCE AND ASTRONOMY. This term is repeated over and over again in the 6th book of this series "**THE DIAMOND PRINCIPLE**" to instruct the public that if they really understood the principles of my work and the true nature of the electrical nature of comets and the solar system at large, then they would soon realize two very important points.

The first is that the Earth has been ravaged in the past by large comets AND THAT THESE ARE THE PLANETS IN FORMATION ... THE PLANET X OBJECTS WILL APPEAR AS LARGE RAGING COMETS IN THE HEAVENS.

The second point is that if we can grab even a small part of this immense electrical energy source of the space in our near Earth environment, then we do not need oil or coal or nuclear power. The economy of the world would change and, in fact, all of the resources we are using up on this planet at a break neck pace would be unnecessary. An additional point is that the ridiculous use of 60-year-old rocket technology to go into space must be replaced with simple electromagnetic propulsion systems to put hundreds of thousands of people into autonomous space stations and begin our birthright to the stars.

Now ask yourself again why it is imperative for the world control freak controllers to keep the public really stupid and fed a regular diet of Tier II fairytale science that could not possibly be true, but appears to have the support of what appears to be the top Ph.D. scientists of the land. If the public woke up one day to see a raging comet filling the sky and started asking the right questions AND GETTING THE RIGHT ANSWERS, the public would soon realize also that the controllers have NO control over that sky monster up there and the public would soon start to ask questions realizing that they have been lied to.

They would wake up to the fact that the controllers have known all along they were feeding the public a line of BS and have spent years preparing themselves for any and all such events while leaving the public unprepared. Instead of preparing, the public has wasted their existence paying ungodly interest rates on over-inflated housing prices and driven their SUVs into the ground racing in rush hour traffic for thirty years, only to have their pension funds stolen by government protected bankers before they get to enjoy their "golden years."

The modern crazy lady talks to aliens Planet X story, as well as dozens of others that permeate the internet, are sponsored to allow the public a taste of insanity to associate with the topic of Planet X and to allow the NASA white hats, as I call them, to come riding in to save the day. The made up stories are so crazy that the NASA Tier II fairytale science actually looks good by comparison. This is the scene presented to the scientifically uneducated public and will be into the foreseeable future.

As I worked through the production of this book, I have come under unbelievable computer cyber attacks from all areas of the government. Keeping my computer world safe and intact was a constant effort, not to mention the constant attacks on my good name that kept appearing on the internet as the public awareness of this new book release hit the public. Their goal was to abscond with and destroy my ability to present my material.

It was on a radio show back in the mid-1990s when I was a guest with a talk show host, Art Bell, that I started to see a level of this game that was not clear to me before. But first you have to understand many of the things I had learned over the years in what is called the "school of hard knocks" or in other words, the real world. Because of my constant pursuit of my work and inventions, I had already had my bleak education in the early 1980s regarding the secret societies that operated the California investment venture capital groups and the international control of the arms industry that kept the USA and Russia on equal footing with respect to arms development.

I knew about the international bankers and their death grip on the governments of the world. I knew about the close knit families that operated in obscurity who controlled the president and congressmen as if they were Jello in a salad bowl. I knew about the CIA drug schemes that operated out of Mena, Arkansas, in the 1990s and the Bush-Clinton connection in the Reagan era Iran Contra “scandal.” I knew that the Vatican had built telescopes in Arizona and in South America and manned them with professional comet astronomers (some of whom I knew from the University of Minnesota) and that they were more than very interested in the coming comets, including Hale Bopp.

I knew that a priest named Father Malachi Martin made public statements that the Vatican had secret connections to the Hubble telescope and directly influenced the workings of NASA, and that the Vatican had great interests in especially comets. I knew that the Russians had a far superior space program than the USA and while we were waiting years for the Apollo moon program to develop the Russians were sending three to ten men DAILY into outer space and that they had a small shuttle craft named KOSMOS that they could place on top of one of the Energy class giant boosters to carry three men into space and return at will. The Russians had over 1200 manned flights in the time during these early years alone.

I knew that the Russians had attempted sending men to the moon long before the USA and these men came back very dead and French fried because of the high radiation van Allen belts that surround Earth, and that they thoroughly knew that NASA did not send men to the moon nor did they know how to breach the radiation belts, let alone land a single engine lunar lander on the irregular surface of the moon.

I knew that as I was personally working with top level scientists in Russia who were learning and using my theoretical work on the electrical nature of the universe, there was a young singer named John Denver who became interested in outer space and, after being rejected from participating in the US space program and NASA (for which he offered to pay to go into space), he paid over a million dollars to the Russian space program and was going to sing his way into outer space and bring the public with him.

Denver died in a mysterious plane crash in the bay outside San Diego and it is said his plane ran out of gas. But, I knew that this was the same deadly technique used by the flight instructors to kill unwanted pilots in the Mena, Arkansas CIA drug running days of the Bush-Clinton Reagan days. I knew that ALL of these issues had large scale cover-ups that were administered from the top, and the international news media was used to present the politically correct story to the unsuspecting public.

When the events of 911 happened, many people denied the government involvement and cover up of these events by saying they could not believe that this could be possible because of the number of officials and other people it would take to keep the story from popping out of the seams somewhere. My answer to this is that the cover-ups were part of business as usual and required no extra work on the part of the staffs and workers and insiders who do these kinds of things every day of the year, albeit on a lesser scale.

Yes, I knew all of these things and more. But regarding my science work and why it had been thoroughly rejected over the years and never allowed to see the light of day I always just thought this was due to egos of the arrogant NASA astronomers and university scientists who had a personal axe to grind. I had witnessed this first hand while I was at Cornell University.

For example, one day I went to visit a member of the Geology department and approached the secretary and requested to talk to the faculty member for my meeting to discuss my work and how this would affect the standard structure of knowledge in the field of planetary geology. The head of the Geology Department was sitting in his office and saw me enter through the open door. He screamed out at the top of his lungs, "No one will get tenure in this department if you talk to McCanney," so everyone, including the professor that I had planned to visit would hear.

Everyone at Cornell knew me at that point and it was clear that my message was not exactly what they would welcome, given the fact that just a few years earlier they had all worked tirelessly to bury the name of Velikovsky. Now they were beset with a tenacious young scientist amongst their very own faculty ranks who was using the data from NASA space probes to challenge their well guarded traditional theories on their own ivory tower turf.

It was in the later part of the 1990s as I was a guest for my annual "Christmas Special" with talk show host, Art Bell, that old Art taught me a lesson that would help me redirect my energy and thought processes on a subtle point regarding the lack of acceptance of my comet and related work. The show was one I had done for a number of years running. I was the guest for the entire show and this particular show we had chosen the topic of my knowledge of the Russian space program. We finally hit on the topic of my science work and Art asked me why my work was not ushered into prominence in the scientific community.

I immediately gave what had become my pat answer, stating that it was the collective and individual egos of the astronomical community and additionally the damage control that would have to occur to recognize officially that all the textbooks and newspapers with NASA information and peer reviewed journals etc, etc, etc would have to be changed and a public admittance that McCanney had been right all these decades, and additionally that Velikovsky had been right. To me this would have been enough to stop any freight train from moving.

But Art, in his wisdom and insight, called me on it and said, "No, there has to be something deeper and more obscure" because all of these things could be dealt with in other ways, including NASA stealing my material and taking credit for it, which in fact happened many times on fine points of my theoretical work. A great example is the outright theft of my Saturn Sweeper Moons paper and work by a Cornell University Professor of Space Science. After all, Stephen J. Gould got away with the outright theft of one of Velikovsky's main concepts and no one batted an eye. So, yes, Art was correct. Everything could have been adjusted and made "good."

Art was also correct that there was another deeper reason for keeping this under the rug. As I said, it was this interview with Art Bell in the later part of the 1990s that made me realize that there were some issues here that someone somewhere did not want the public to know about or understand. My topics breached too many openings into the control network that kept the public scientifically stupid and kept the world leaders firmly in control.

The flat Earth science that was pawned off on the public, as well as the structure that protected it in Tier II and III levels of science, was very important to nothing more than crowd control on a worldwide scale. Simply put, if the public were to realize en masse all of my work was true and they looked up into the heavens one day to see one of the Great Comets raging in the daylight sky from horizon to horizon and dashing electrical discharges to Earth or one of the other planets, the glue of international control would come un-glued in a New York second, and everyone on this planet would realize overnight all the uncomfortable implications of this stark realization. The wizard behind the curtain would be unveiled, and Dorothy and Todo would be free to go home.

In July of 1999 Father Malachi Martin was murdered and we searched for answers. He was instrumental in exposing the Jesuit-run Vatican and the underpinnings of their involvement in the international business of secret astronomy, specifically the astronomy of comets. Father Martin, because of his inside knowledge, knew the question that came to NASA regarding the comet Hale Bopp was "Is this Wormwood?" This in itself says it all. The Vatican knew from historical records that the coming end of the world would be at the hands of a Great Comet and they thought that possibly Hale Bopp was it. It is little known that when Hale Bopp was first discovered it was on a direct collision course with Earth. The best estimate was that they could not calculate if it would hit or just be a near miss.

The nucleus of Hale Bopp was big and very massive, at least the size of our moon, and it held a companion. In astronomical terms, this was unheard of since comets were supposed to be little puffy snowballs with no gravitational field to speak of and certainly none that would hold a companion in orbit. I have thoroughly discussed the Hale Bopp companion issues on my radio shows and other books, so will not reiterate here. But basically, NASA publicly denied it and ruined the careers of many people who attempted to report the presence of the companion. Thirty-seven people in California, in a group called Heaven's Gate, allegedly committed suicide over the companion's presence and NASA produced photos that "proved there was no companion." Art Bell was publicly accused of being responsible for this alleged mass suicide because he posted the photos from various sources showing the companions.

After all the negation and international fanfare, at the 1997 Canary Islands International Conference on Hale Bopp (sponsored by the European Space Organization ESO and heavily attended by NASA scientists and astronomers, including the ones who produced the press photos of the comet with NO companions), the first topic on the roster was the confirmation of the companion of Hale Bopp. In spite of this, NASA never made the public announcement regarding this contradiction and the official story today remains that there was no companion.

This is a complex web of misinformation and cover-ups, which go to the top of the command, but the reason for this is the same reason why there is no real information given to the information hungry public regarding the true nature of Planet X. Note that the Vatican was fully aware that incoming planets take the form of large comets from historical records. Note that all of their high investment telescopes are specifically designed to view comets and are of absolutely no use to view other deep space objects, such as galaxies and quasars. They are strictly to view comets. They are obviously deeply concerned about the effects of these objects on Earth, yet the public is fed a diet of "fluffy little snowballs" that cannot hurt a celestial flea.

In the mid 1990s, as the comet Hale Bopp was lighting up the day and night sky, a misinformation campaign was originated with ideas that could only have come from professional astronomers. A woman who was involved in programming what was known as PROMIS software and her estranged husband, who was owner of the company, spearheaded a CIA sponsored program of deceit regarding the topic of

Planet X. As I have stated before, in any well designed misinformation program the controllers control both sides of the dialogue. To have real crazies out there would not give predictable results. And so it was with this program to defame the concept of Planet X (to make it into a topic only crazy people would consider).

A crazy lady who claimed she channeled with alleged alien "friends" was construed. The aliens would tell her all about Planet X (only she could hear them in her brain, goes the story) and she would relay this to the public. How anyone like this ever got to international fame and placed on all the top radio shows is purely the result of a well orchestrated high powered advertising and promotional campaign.

On May 15, 2003 an alleged bright orange Planet X was to rise over the horizon and crash into and destroy Earth. Years of planned misinformation accompanied this campaign, complete with large scale radio show coverage on top international radio shows. Mass suicide events were planned and international press coverage peaked to show the crazy lady's predictions fall flat on their faces. Then afterwards everyone could laugh and point to all those crazy Planet X people. They wanted to make this the craziest story that anyone had seen.

In spite of the fact that I had to dedicate much of my personal radio show to dispelling the crazy information and try to direct traffic in the right direction, the misinformation mill consistently tried to associate my work with this batch of well planned misinformation. They wanted to end the topic of Planet X in the public eye, once and for all. The only problem was that this effort somewhat backfired on NASA and the planners. A few days before the alleged end of the world I discovered the true meaning of the May 15, 2003 date, and this was something that only would have been know about by esoteric group of astronomers. On this date would be a strange ORANGE FULL MOON that would first rise over the horizon in Japan, then the west coast of the USA, and then still later over Europe, as reports of mass suicides came in from these locations.

I got on the phone and directly called NASA, stating that if they did not immediately use their power of the press to negate this foolishness and announce that the real thing people would be seeing would be this unusual full moon, that they would be as responsible for any mass suicides as the lunatics who were promoting this nonsense. The mass suicides were squelched and the story of mass suicides by cultist Planet X following the alien channeling crazy lady, never materialized. Overnight eight years of planning and hoping to kill the topic of Planet X fizzled, and the end result was far less than hoped for by what we have fondly come to call "the disinformation crew," which includes professional astronomers working behind the scenes at NASA.

More recently there was another wave of Planet X hysteria during the recent 2012 end-of-the-world Mayan Calendar misinformation campaign. Once again it is questionable if this really had the intended effect of turning off the public to the topic of Planet X. Comedians and even some TV commercials got good mileage out of this and many in the public have been turned off to "preppers" who prepare for future unknown events.

I have to always make it clear that I do not predict dates of celestial disasters, although my detractors are constantly trying to prod me into it. As this book moved towards public distribution, numerous misinformation campaigns were launched in the public media to try to make it look like I was predicting the end of the world and cause mass suicides (like allegedly happened with Hale Bopp). But, with each such event the instigators have been exposed and eventually the fooled public wakes up and realizes that there is a concerted effort to ruin my good name and place me in a league with the crazy people who claim to talk to aliens and the like.

This is a result, actually, of my success in educating the public. This would not be happening if I were failing in my attempts to reach the public eye. All this occurs with a budget of really nothing, only using my mind and ability to release information directly to the public, bypassing the standard controlled news media. I do not rely on people believing me because of a badge or government official title. I depend on the public understanding my work on a level they can visualize and my ability to explain complex scientific details on many levels.

In summary, the main points to remember regarding the topic of Planet X and my work in general is as follows:

1) Comets are not dirty snowballs

2) Outer space is electrically active and is driven by the fusion at the solar surface (solar fusion is not at the core)

3) All the planets and comets from the largest to the smallest piece of dust will discharge the solar capacitor, causing subtle but observable electrical effects

4) Comets, according to my Plasma Discharge Comet Model, show that comet nuclei can be any size from small dust grains to objects larger than the planet Jupiter

5) Most comets are harmless and at best provide us with a pleasant visual display in the night sky (and sometimes the daytime sky) – very few are dangerous

6) All of the planets and moons are also discharging the solar capacitor to a lesser degree than comets on elliptical orbits around the sun (as explained in my previous books and lectures) and this is a major cause of planetary weather

7) Electrical effects in the solar system are abundant and must be understood in context, since gravity is still the dominant force governing the cosmos

8) Comets can affect Earth and the other planets by ACTION AT A DISTANCE, including gravitational – electrical – magnetic and other plasma effects

9) The tail drag of the comet, along with orbital changes due to gravitational interactions with existing planets, cause some incoming comets to be adopted into the solar system as new members, including moons and planets

10) Planet X objects will appear as large comets. Not all will cause direct devastation to the Earth or the other planets. Each comet must be analyzed individually.

11) There is a tremendous effort to keep this information from the public because the results of waking up en masse would cause an immediate change to the world power structure.

THE END –

OR IS THIS JUST THE BEGINNING?

ADDENDUM
THE FUTURE OF MAN & EARTH

Are we destined to repeat the errors of Man (the collective) in the past or can we redirect the insanity of our current plight to save the species? In this chapter we will take the major problems on Planet Earth and discuss them one by one. As it stands, Man & Earth are on a collision course for mass destruction. If left unchanged, we will eventually face complete destruction from outer space at the hands of a large passing comet. Prior to that, we may destroy ourselves, and that seems to be closer to reality every day. How Man & Earth escaped complete destruction from celestial events over the past 10,000 years is pure chance. We dodged many speeding bullets, literally, and then slowly recovered. Central to these issues are understandings of some basic principles.

Government systems are very organized from higher up and are universally designed and controlled by a small group of people. It is the result of inevitable confluence of wealth and power. In addition to out-of-control governments, billionaires use their money for social engineering (some good and some very bad). Groups span the globe with agendas that are tugging at the very fiber of civilization. The individual is lost in this sea of forces battering them in all directions.

Education and every other institution have been infiltrated so that no one can trust anything any more. It is like being adrift in a dingy on the ocean with ten hurricanes coming from all directions. The only real hope for some is isolation and literally returning to the cave ages, or at least the pioneer days, on a quest for survival. The evening news is controlled and NEVER discusses these issues except to poke fun at the "preppers" who are made out to be a bunch of crazed lunatic cults.

With that as a realistic assessment of the world today, here are the eleven major issues we face with solutions for each that must be implemented to return to a level of sanity on planet Earth. It is my firm belief that if we solve the first two (Energy and Water), all of the rest will automatically be solved. In fact, if we solve the energy problem, the second problem—WATER—will also be solved. But solving the first ten problems will not save the human race from our inevitable mass extinction. Only our understanding that we have to get large populations (millions of people) into many autonomous space stations will ultimately save the human race.

Landing three men on the Moon or Mars in the near future WILL NOT solve anything, except boost the egos of those involved. From the very beginning of the private space ventures, I predicted that the "powers that be" will not allow private ventures into space and will sabotage them as they get close to succeeding with human space programs. I have been correct on this count 100%. The only ones who will be allowed to succeed will be those selling out to the already existing power structure.

Remember that outer space is deemed to be the private playground of the world banking cartel. All other aspects of daily life are to control people to keep their feet firmly planted on good old Terra Firma. The banking cartel's private police force (the CIA) controls NASA, which is the Tier 2 smoke screen front agency whose entire purpose is to make the public believe there is a space program going on. NASA has never been about letting the public into space. Since their inception, they have essentially accomplished nothing. In fact, NASA's entire purpose has been to prevent us from going into space. As the Tier 2 rockets are launched, go cheer them on and keep sending those dollars for the NASA annual budget, as each passing year brings us closer to a celestial mass extinction event.

Now, to discuss the ten top world problems, with the eleventh being added relative to saving the human genome from extinction. You will notice that Global Climate Change is not on the list. This dog-and-pony show was begun by Carl Sagan's "Greenhouse Effect," which does not work on Venus (where it all started), let alone on Earth.

The biggest problem is that a real greenhouse has a roof that prevents the heated air from physically leaving, whereas our atmosphere does not. Every day, the heat energy from 80% of all fuel burned goes directly into the atmosphere. The earth sheds this daily with no problem at all. This is literally billions of times more heat than the twenty year expectations of a Greenhouse Effect. It has never been anything other than bad science and a political football to control humans, and especially WATER. Relative to world problems, it would not be included in even a top 100 list of world problems. Its use as a political football would rank it near the number 1 spot in man made boogeyman problems. Now, for the top ten problems with solutions for the top three—Energy, Water, and Food—as well as the tenth issue—Population—and the eleventh issue of saving human DNA.

1) Energy

a. Problem: We use far too much energy. First world countries build energy inefficient houses and run economies using energy as if it grew on trees. The "indicators" that are used to drive us to make the same number of cars and building houses each year is a false economy. It cannot last. The days before the Noah's Flood are said to have been the same with the human population using the world's resources until they were all gone. Energy became a mercantile commodity in the late 1800s. As with any commodity, ownership has become monopolized. Anything that would break that monopoly is prevented. Prior to that, all progress was based on human and animal labor.

The energy industry is probably the most corrupt entity on the planet. Every ten years there is a charade promulgated that we are running out of energy resources. It simply is not true. The real issue is that the easily accessible energy sources are being used up. Energy will be harder to "harvest" in the future, but it is not in short supply. The "alternative energy" industry, which is supposed to save us, is very corrupt.

GE and Siemens, the two largest manufacturers of wind energy equipment, have been marketing a "solution" (the 3-blade wind turbines) that simply do not work. The entire purpose was to produce a product that did not work, suck all the existing money out of the governments of the world, and when everyone woke up to the fact that these will never solve the energy problem, the cry would go out to return to the GE and Siemens core business of centralized power.

Electric cars (and future hydrogen H2 cars) must rely on central power company electricity to create the "fuel" and are more environmentally damaging and less efficient than just burning the "fossil" or nuclear fuel in the first place. Nuclear power is the most horrific polluting energy source on the planet and never should have been approved. Some day the spent fuel will literally kill off all life on the planet as the casks that store it break open.

b. Solution: We must cut energy usage by 90% but with the caveat that is has to be done slowly. Raising fuel prices will be the only method of forcing the public to respond. Nuclear and Coal as sources of fuel must be eliminated (but again with the caveat that it must take place over a minimum ten year period). Petroleum also must be eliminated. The entire problem being pushed by politicians today is that they are trying to replace stable fuel sources with "alternative energy" sources that do not work.

Solar will never solve the world energy problem because, simply put, it will always require huge installations of batteries or other form of energy storage. The wind is the best renewable energy source. It blows day and night, from the north pole to south pole, when it is raining and sunny all year long. It can never be depleted, because every day the wind is renewed by electrical currents from the solar capacitor. We just need an efficient device to extract the energy from the wind. The JMCC WING® Generator is the solution. 95% of our energy needs can be met by direct generation of electricity from wind.

Direct analysis of the wind shows that large areas (countries or states) have a constant supply of wind energy (if it is not blowing in one location it is blowing in another).

The failed 3-blade wind turbine technology has given the false impression that wind does not work. That false impression was intentional, as already noted. It is possible to build out 150% of our energy requirement with the "WING," which also eliminates the necessity of batteries. A single 80% efficient 200megawatt JMCC WING® Generator, in a good wind location, can replace 400 2-megawatt 3-blade wind turbines operating at industry average of 20%. Using just 5% of the energy from Natural Gas in the central power companies, the country can literally move to complete independence from the big three—Coal, Nuclear, and petroleum—and move to 95% wind energy within ten years. For transportation, Green H2 hydrogen from electrolysis will replace petroleum from transportation. A 200 mega-watt JMCC WING® Generator system placed on an ocean going barge can be designed to produce 100,000 kg of Green H2 per day per unit at a cost of $2 per kg, and without government subsidies.

The future of world energy is Natural Gas, direct electrical energy from 95% JMCC WING Generator and 5% Natural Gas, and Green H2 using the wind for transportation in place of petroleum. Dry H2 storage technologies will be used (not pressurized H2). We will recycle the waste water of burning H2 that has already been purified.

2) Water

a. Problem: We live on a water planet, yet water is one of the scarcest commodities. This is by design. There are many concepts that are totally false, including many included in the Global Climate Change conundrum. The earth managed much larger herds of animals than exist today (the bison herds of the 1800s) yet people believe that animal outgassing is a modern problem. "Climate Change," which evolved from "Global Warming," which evolved from "The Greenhouse Effect" have never been anything more than bandwagon "science" to get people to give up their inalienable right to water. See my book "WATER" for more details.

b. Solution: As with energy, we do not have a water shortage problem; we have a water management problem from local to national to international centers of control. The driest desert areas on Earth receive more than enough rain to support large populations locally. Earth's most abundant surface element is H2O WATER. Who could convince you that there is a shortage? We have the technology to remove salt from seawater … it simply takes energy.

Using Reverse Osmosis (RO) at thirty cents per kilowatt hour of energy (typical cost on remote islands), it costs just two cents per 100 liters of water. Between rain water storage and RO, 80% of the world population can have fresh water with the other 20% using underground or surface sources. One fact is that water must be procured locally. Transporting water is neither efficient nor economically practical. A simple calculation is as follows:

Take Sierra Leone a small African country with population 8 million people. At twenty gallons of water per day per person (considered to be minimum for survival), this small country will use 160 million gallons of water per day. By comparison, first world countries use about 200 gallons per person per day. The largest oil transport tankers in the world can hold ten million gallons, so it would take the volume of sixteen of the world's largest oil tankers PER DAY to meet this small country's minimum water requirement.

Assuming we could convert oil tankers to carry water and assuming they would be taking fresh water from the Greenland Ice source, and assuming it takes two days to load and two days to offload each ship, plus six weeks round trip, for a total of 46 days per ship to make the round trip, it would take 46 x 16 = 736 major oil tankers operating 24/7/365 to supply little Sierra Leone with its minimal supply of water. This does not include the fuel oil needed to move these ships.

As you can see, moving water is not possible as a worldwide solution. Water must be procured locally. This is why insane mantras of Global Climate Changers, such as stopping cows from drinking water in Kansas, will provide water to poor people in India are simply fantastic misperceptions not based on reality. Water catchment with filtration, supplemented with RO from seawater is the least expensive and most effective source of water for about 80% of the world's population.

3) Food - Famine (both real and man made)

a. Problem: As with energy and water, the world food shortage problem is man made. Since these are needed on a daily basis, the easiest way to control large populations is by control of these three essentials. Another calculation tells the entire story. Imagine an organic farm the size of the state of Iowa (thirty-five million acres) growing organic non-GMO vegetables and fruits, with chickens and fish as the primary protein sources whose waste is used for fertilizer. Standard estimates are that an acre of organic farm can support 100 people (with one crop per year). This small area would support 50% of the world's population. If you chose a location like California that has multiple crops per year, it would produce enough food to feed the entire world ... COMFORTABLY. Now look at the entire world agricultural regions. They collectively have an estimated 4.5 billion acres of land under cultivation. At 100 people per acre (assuming a single growing season per year) this area could support 450 billion people, or about 65 times the current world population. So why is there a food shortage anywhere?

b. Solution: The ownership of food, as with energy and water, are behind the alleged shortages. Recent wars are blamed for food shortages but, even with that, there is still more than abundant food production. As with water, it is far easier to produce food locally and is very expensive to ship long distances. Even far northern and southern areas of earth have very good farming seasons. Fruits and vegetables in Alaska grow faster and far larger than in other regions. Aeroponic Grow Towers, which can raise fifty heads of lettuce in the space of one planted in the dirt, use 95% less water with plants maturing in 1/3 the time of traditional farming.

In outer space this will be used to recycle CO_2 back to O_2, and provide autonomous space stations with food. Aeroponics can be enclosed in CO_2 rich atmospheres to grow larger and faster, using grow lights. One can house these facilities in abandoned shopping malls, which will then have restaurants and food stores with food grown locally. This is the future of food and agriculture in urban settings. Returning to the main topic of food and famine, there is no food shortage, there is a food management and ownership problem.

The following worldwide problems, 4 to 9, are noted for later discussion.

4) Environment (both man made and natural)
5) Poverty
6) Terrorism & War
7) Disease
8) Education
9) Democracy

The 10th and 11th problems below have both a Problem and Solution discussion.

10) Population

a. Problem: Is the world really over populated?

b. Solution: If you took all of the estimated seven-billion people in the world and gave each a small apartment, they would all fit into the Grand Canyon. The world populations are actually decreasing at an alarming rate.

Married couples in first world countries are waiting longer and having fewer children, with the average population decreasing. China is trying to import women to have more babies as their population declines. A single celestial event or man made event could wipe out a large sector of the worldwide population in a single day. With supply chains disrupted (by natural or man made causes), even more people will die. There is a myth that the world is over populated.

I am adding the 11th and most important issue facing the entire human population and that is to prevent the ultimate demise of the human genome. If you look at my life work and inventions, they solve the eleven essential problems. Now maybe you will understand why there is such an overbearing effort behind the scenes to stall my ability to reach the public at every pass. I identify the real problems and present real solutions. Scientists are supposed to be quiet mouse-like creatures in the corner, filling out their grant proposals and publishing in a never ending collage of mindless "science." I have never fit into that mold.

11) Saving the human DNA from extinction - Outer Space and Saving Human Genome

a. Problem: No one talks about this topic, including NASA and even those spending their personal billions on private space programs. Most revolve around sending small teams to Mars or space tourism for an expensive joy ride.

One major point of this book is to point out the controlled opposition "an asteroid hit Earth" theories that allege a catastrophic single event took place so long ago that no one has the feeling of "recent" and therefore "imminent" as far as we are concerned. It cannot happen to us, is the result. That is the scope of "allowed" controlled opposition.

The reality is that there have been no less than five major earth wide disasters in the past 10,000 years with the most recent (and possibly the worst event) just 1500 B.C. This was caused by a large passing comet, as was the Noah's flood event some thousands of years earlier. Man as a species dodged these celestial bullets and came back out of the ashes. If we continue to live in the delusional world that this will not happen again, then the human race will lose in this deadly celestial game of Russian roulette.

b. Solution: I call them "bubble societies." These are small dedicated groups that put their efforts and resources into building the equivalent of the Noah's Ark in the form of autonomous space stations, capable of holding 100,000 people or more. This is the number of people it will take to maintain a diverse genetic pool for maintaining the human race. Modern venture capital and government space programs do not fit this business model, as they all need some kind of return on investment. Autonomous space stations do not come back, so there is no ROI. The insane world leaders consider space to be their private playground; however, inbreeding within select small genetic blood lines has been the downfall of every monarchy throughout history.

The main theme in my book "The Diamond Principe" is on discussions of the human condition and man's ability to psychologically gain a foothold in outer space. In that book I discuss the runway to space hotel tube launch system, which also has details on my web page. The idea is that landing on other celestial objects and trying to establish colonies, like on the Moon or Mars, creates a much more difficult technical set of problems than creating rotating autonomous space stations with artificial gravity. Man's tranquil environment could change at any time with the only way of preserving our DNA being an escape in large autonomous space stations.

Does man have the capacity to understand this? If so, does Man (the collective) have the wherewithal to take the necessary steps to become space fairing in large numbers? We do not have to travel to the nearest star, as Earth will recover. We just have to wait a period of time before returning as Earth itself recovers. Has Man ever accomplished this before? Is Earth a test bed to see if man has "The Right Stuff" to save his DNA? Or are we doomed to repeat history and, at best, relay to future generations our inevitable destruction by way of Myths Magic and Mayhem?

THE END

NOTE: The introduction of this book is reprinted on the following pages. Read it again to digest the intent of this book.

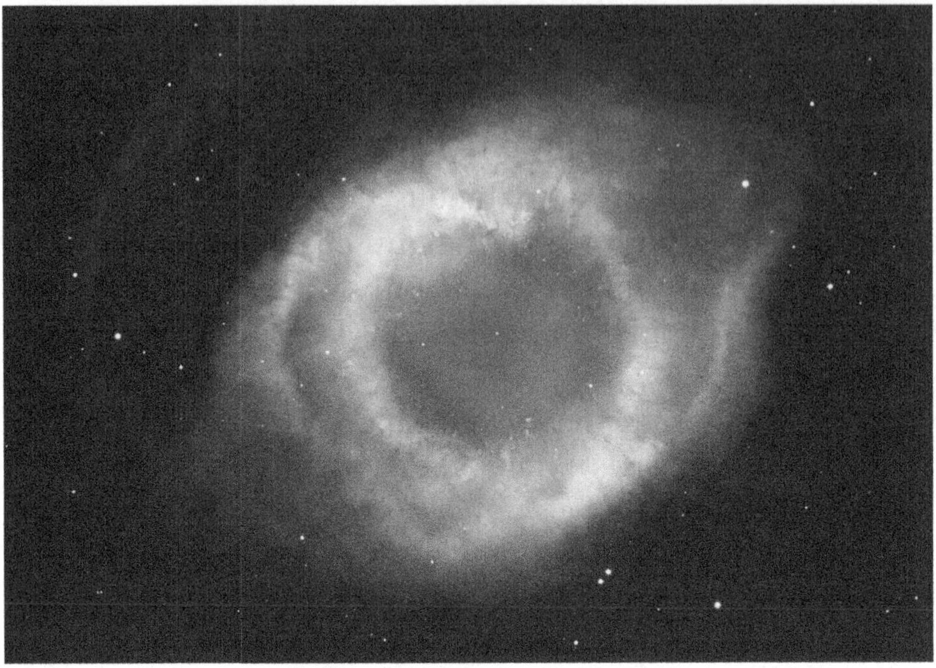

INTRODUCTION

The above photo is of one of millions of stellar "nebular ion clouds" surrounding stars that cannot be explained by standard astronomical theory. If you could travel a short distance out of our solar system and look back, you would see our sun to be very similar. It is a cloud of dust and gases held back by a positively charged proton wind that emanates from the fusion based star (what we call the "solar wind"). This creates an electrical capacitor that is instrumental in causing a myriad of effects and additionally provides the energy for comets to become planets (as explained in my Plasma Discharge Comet Model) in a complex cosmic game of celestial billiards. This book discusses how these physical processes have drastically affected the History of Man and Earth, leaving us a view through the eyes of the ancients of Myths Magic and Mayhem.

To put this photo into perspective, the center dot would be our sun and planets out to about Earth. The inner part of the nebular ion cloud would be about where the planet Pluto orbits the sun. The region with brightly colored nebula would be what astronomers call "The Kuiper Belt" where tens of thousands of forming planets have been discovered in recent years. You can see the large comets forming planets. One by one, Jupiter and our Sun (a twin star) captured all of the planets and moons into the solar system by a gravitational process known as "capture." All of the existing planets initiated in the form of large planetary sized comets. My science describes the physics of these processes in my "Plasma Discharge Comet Model."

This model shows large comets growing into planets as the tail material is pulled ***into*** the comet nucleus by powerful electrical forces. It is this same influx of matter that causes what I term "tail drag." This causes the newly formed planets elliptical orbits to become circularized and fall into the orbits around the sun that we see today. During the planet capture and formation processes, the huge comets can cause great devastation to the existing planets like Earth (sometimes during multiple passages). They do not collide with the existing planets, but pass nearby. This scenario explains the many earth effects not explained by a single small comet impact whose energy is absorbed locally.

"Action at a Distance" is the term I coined for the

thousands of effects that can take place when a large planet-sized comet nucleus and its electromagnetic plasma sheath passes near Earth or other planets like Mars. The ancients tell us through their writings and petroglyphs about their experiences with such events. Man's development and Earth itself have been greatly altered by these events. My prior books go into great detail about the Earth Changes and effects, which will only be briefly summarized in this book.

They include the following: Pole shifts, mountain building, rapid plate movements, massive earthquakes, worldwide volcanic eruptions, tsunamis in both the oceans and land, blocking sunlight causing rapid radical cooling, glaciation, ice ages, severe weather, influx of materials from the comet tail including water, hydrocarbons (oil what the ancients called naphtha), meteors, dust with high concentrations of minerals, nano-diamonds (which are now known to form to the sunward side of comets), worldwide auroras and upper atmospheric electrical effects, huge vibrational sounds (the ancients called trumpeting), electromagnetic effects, including comet to planet discharges, and the list goes on.

Large influxes of space dust can cover existing man made structures with dust and sand in a single day long passage. This makes more sense for sites such as Gobekli Tepe and hundreds more, than to believe that the inhabitants buried their own thriving cities.

This leaves the earth reeling for decades and centuries to come. The ancients who survived such events pass down their experiences attributing them by way of Myths Magic and Mayhem. A single small comet strike cannot explain these same effects, which are almost universally observed in the geological record. The average groups studying human history today, which includes professional astronomers, archaeologists and independent researchers, are so affixed to the "an asteroid or comet hit it" concept that they have no idea that it does not explain the myriad of effects we observe in the hundreds of locations. Additionally, the earth has experienced many hundreds of such events, not just one or two.

My book "**Principia Meteorologia - The Physics of Sun Earth Weather**" (2004) details the effects of the passage of a large comet, not only on planet Earth but other planets such as Mars. Mars shows that it recently had an atmosphere and oceans that were removed as a large comet passed by and removed these as observed by the ancients. The original book of this library series, "**Planet X Comets and Earth Changes**" (2000) as well as the follow on books, CDs, DVDs, and Lecture Series "**Surviving Planet X Passage**," "**Atlantis to Tesla – The Kolbrin Connection**," "**COMETS**," my 18 Hour MP3 Lecture Series "**The Physics of Ancient Celestial Disasters**" and the 180 Hour Lecture Series "**Science Home Education Package**" further detail the history of man and Earth. The current book is a summary of the various aspects of these works and is being released at a time when the public is finally waking up to the reality of multiple recent massive earth wide disasters and dispelling the slow evolution history promulgated by NASA and standard astronomy.

This brings us to some major conclusions. Man's history has had MANY Earth-shattering events, including near mass extinctions. This book describes these events as related to the physics of comets and solar system formation. Earth has been assaulted and re-layered many times in its history. That is why the more we dig, the more we find unknown ancient ruins of lost civilizations, scorched and buried in the aftermath. The true age of planet Earth and the other planets of the solar system cannot be known, because the surfaces have been drastically altered many times. Dating material found on the surface as has been done will not produce correct date results.

Researchers who still believe that extinction event Earth Changes were caused by small celestial objects colliding with Earth, must see that small asteroids or comets colliding with Earth do not explain the worldwide effects we see. For example, using my Plasma Discharge Comet Model and laboratory experiments, micro diamonds form to the sunward side of the plasma discharge comets, just as diamonds form in the laboratory vacuum chamber with carbon rich gases in an electron current. During a close passage of Earth by a large comet, uncountable numbers of micro diamonds rain in along with other forms of cosmic pollution.

The idea being proposed today by researchers (both from the archaeology and science astronomy worlds), that a small comet hitting Earth would create millions of micro diamonds, is simply incorrect. This book addresses these issues in detail. Two points are clear. First, devastation at the hands of a large passing comet has happened many times in Man and Earth's history and secondly, buried under repeated layers (not just one or two) are the remnants of many deceased civilizations of man. The history of Man on planet Earth may go back millions of years or even much more. The Real History of Man and Earth are far different than what you have been taught. This history is filled with Myths Magic and Mayhem.

The main reason for releasing this book now is because there is so much controlled misinformation emerging now regarding two events called the "Younger Dryas Event," which researchers claim occurred some 12,800 and 11,500 years ago. They claim that a small comet hit Earth and caused all kinds of unrelated events. One great problem is that dating techniques are more of a bandwagon consensus rather than based on sound science. Unfortunately, many researchers are jumping on this bandwagon. Here are some main points. First of all, this has been thoroughly presented before many times starting in 1979 by two astronomers Clube and Napier, who envisioned multiple impacts.

They sold a lot of books and gained fame only to have this fall by the wayside. Secondly, and most importantly, a single comet hitting Earth is assumed because it is politically correct (it agrees with NASA propaganda that comets would be harmless little snowballs) and thirdly, none of the worldwide effects they are claiming could be the result of a small comet impact. A side issue is that we know there were numerous devastating events much more recently than the out-of-sight-out-of-mind 12,000 year dating scheme. And lastly, they cannot find the crater associated with such an event that would certainly be very visible on Earth's surface. Every ten years this same scenario rises out of the ashes. The problem is that the real reasons for what we observe in the archaeological record have been established as being caused by large comets passing by Earth by "Action at a Distance" effects that do, in fact, explain what the ancients describe (the huge comets in the sky that attack Earth and pass by some to return again another day).

Decades ago, there was a group claiming that the dinosaurs went extinct in a single comet strike of Earth. They claimed that the crater was found in the northern region of the Yucatan peninsula in the Gulf of Mexico. The fanfare was heard worldwide and is, unfortunately, still repeated as some kind of established fact. This was finally laid to rest when diligent researchers studied the animal life below and on top of this crater (which was, in fact, caused by such an asteroid or comet impact). They found that the dinosaurs did not even die off locally after the impact, let alone worldwide.

Why this debunked scenario is gaining air again speaks of a giant disinformation campaign. We are finding out that some people are receiving money under the table to "continue their research." It is great for their book sales, speaking tours, and the fame and all that goes with it. Below the surface, however, this speaks of a well funded misinformation program to popularize a completely improbable scenario while squelching the real story from the public. We are already seeing

Netflix and History Channel specials and blanket coverage in the international fake news media. This is clear evidence of a massive, well coordinated distortion that is designed to divert the public from real science, real archaeology and the true history of Man and Earth. Today, man is at a critical crossroads and researchers who present mountains of good information but come to completely false conclusions (while greatly profiting) are useful pawns in this deadly game of science and history distortion.

These stories gain momentum and it becomes fashionable to jump on board. There may even be a Nobel Prize in it for some of the more aggressive Ivory Tower supporters of this movement. Objections are drowned out in the rush to fame and blanket coverage in the controlled news media. Those partaking in this charade love the attention while always believing they have hit the mark and are changing the narrative for the advancement of the human race. The easiest way to control most people is via egos.

I am releasing this book now in an attempt to reign in on this disinformation campaign (just as I have done in the past with the bogus CIA Planet X intel program, the Millennium Bug scare, the 2012 Mayan Calendar baloney, and countless other CIA misinformation programs). Since I was a young person living in Mexico City (early 1970s), studying both archaeology and the physics of solar system formation, I have been involved in these issues.

I have been involved in this long before most of the current crop of "researchers" were born. I have seen misinformation come and go, and also witnessed the distortion of history as academicians, archaeologists, physicists and astronomers, geologists and many other professionals amassed volumes of information but used "accepted science" to come to completely incorrect conclusions. To gain pubic view in the controlled media, one has to be politically correct, relative to standard astronomy claims of how the solar system works and its time frame based on slow cosmic and biological Darwinian Evolution.

The biggest mistake made by all is to put dates on events. Dating is nearly impossible, yet the very first thing most researchers do is to put a date on a place or event. They argue against current dating structures, but immediately enter their own dates. This only serves to confound the real history of man. This book, hopefully, will put some sanity into the current fray of misinformation hitting the public in the face today.

Today independent researchers are breaking away from the archaic ideas of archaeology, stating what should have been obvious from the very beginning. Many early archaeologists had the concepts correct (that there were other far more ancient and advanced civilizations on earth that created the amazing stone work). The interpretation of ancient architecture and building techniques are not explained with standard explanations and one has to make the connection to a much deeper past and earlier technologically advanced civilizations. These same researchers, unfortunately, then cling to archaic outdated astronomical concepts in attempts to explain the celestial influences that are visible at the countless ancient archaeological sites. It is time to also step away from outdated astronomical theories and recognize that our universities, museums and planetariums are filled with "textbook repeaters" still propagating ideas that could not possibly be true.

I have a suggestion to all researchers and interested parties. First, let us remove the dating structures both old and new because, literally, we are just beginning to unravel the true story. Secondly, let all involved in this renaissance of archaeological discovery come to the conclusion that standard astronomical theories (as promulgated by Carl Sagan in his over rated COSMOS series) are completely erroneous and a completely new set of astronomical concepts must take their place based on my extensive reconstruction of the workings of the cosmos and the Plasma Discharge Comet Model.

Below is another example of what I call in my theoretical work the "Nebular Ion Cloud." This is an angle view of the plane of that planetary system. Here, too, you see the large comets forming planets (with their Plasma Tails pointing away from the central star). This region is the equivalent of our Kuiper Belt where thousands of new planets are forming.

When the New Horizons space craft arrived at the planet Pluto, which is on the inner edge of our Nebular Ion Cloud, it saw the comet tails of both Pluto and its companion Charon. They also show photos of the comet tail material falling into the planet as ammonia and other nitrogen compound "snow" but, as with all data, NASA could not explain it so they ignored it. My book library and lecture series, written for the general public and professionals alike, are available and must be read to put this book into perspective. They detail my theoretical physics developed 1979 to 1981 as I was on the faculty of both the Physics and Mathematics Departments of Cornell University.

This replaces the NASA dirty snowball comet model and other concepts of astronomy that have been thoroughly disproven by space craft data. My books and lectures library make a wonderful gift for youth, their parents, or for serious researchers who want to know the real story of the cosmos that we live in. Follow the link below for more details.

https://www.jmccanneyscience.com/

ABOUT THE AUTHOR

Suffice it to say that the author is known internationally for his contributions in many venues and in many different fields of endeavor. All of the other books in this series contain extensive information and his bio. Since you are required to read the other books to put this book into perspective the "About the Author" section is omitted here. For those unfamiliar with the extensive and diverse background, the following is how "The Professor" is typically introduced on the many international radio and television programs where he has been expert presenter and guest.

James McCanney, M.S. Physics-Mathematics hosts "The James McCanney Science Hour - At the Crossroads" which airs worldwide from WWCR Shortwave Station Nashville, TN (since 2000).

McCanney is a favorite guest on national and international radio and TV venues. He has appeared in

History Channel Specials and on prime time Russian and Japanese TV.

Contact information: Jmccanney@usinternet.com
Home Page: https://www.jmccanneyscience.com
Short URL: jmccsci.com

Professor James McCanney has taught the following mathematics courses at University level in addition to Physics, Computer Science and Astronomy; Abstract Algebra, Linear Algebra, Matrix Algebra, Probability and Statistics, Statistics for Computing, Mathematical Logic, Theory of Numbers, Calculus I, II and III, Engineering Math I and II, Advanced Topics in Geometry and Topology. He has spent another 25 years in the telecommunications industry with expertise in cybersecurity, encryption, and communication protocols. Much of this was accomplished in multi-lingual settings, having worked in the USA, Latin America, and with high-level Russian scientists at the University of Novosibirsk. He has published peer reviewed papers in astrophysics journals.

He has presented his research at international conferences and is a regular presenter at American Geophysical Union meetings. He has also lectured at Los Alamos National Laboratories, the Air-Space/America International Air show and International Electric Propulsion conferences. Understanding his background is important in placing his books and teachings into perspective.

He is inventor of the patented JMCC WING Generator – a wind energy device that will provide energy for farms and ranches, remote islands, and small communities as well as large cities … ushering in "The Jet Age of Wind Energy."

He entered the elite group of XPRIZE recipients being awarded the 2018 "XPRIZE Technology Innovation Achievement Award" for his work with renewable energy and water from the air technologies being selected over 98 international corporations. His interest areas are Astronomy, Astrophysics, Space Science and exploration, Electromagnetic Propulsion Systems, Energy, Water, Food, Shelters, Prime Numbers and Education.

OTHER BOOKS and INFORMATION

Professor James M. Mccanney, M.S. Physics

www.jmccsci.com for eBooks and hard copy books

BOOKS AND EBOOKS (some available also in Spanish)

Planet X Comets and Earth Changes

Surviving Planet X Passage

Atlantis to Tesla–The Kolbrin Connection

Principia Meteorologia – The Physics of Sun Earth Weather

Calculate Primes

The Diamond Principle

COMETS

WATER

Breaking RSA Codes for Fun and Profit

McCanney WING Generator – World Energy Project

The Nazca and Palpa Lines of Peru –
Mystery Solved – White Paper

Principles of Prime Numbers – Volume 1 of 3

Planet X Update 2020 – The Real History of Planet X

CDs AND DVDs

2 DVD set – Historical video lectures re-mastered to DVD
DVD lectures series – DVD # 3 – Re-Writing Cosmology

DVD – What the Japanese Want to Know About Planet X

eDVD – University Level Astronomy Astrophysics Space
Science Course

eDVD – 20 Yr Anniv. Hale Bopp Harrington Expedition Planet X Update

CD – 110 high resolution aerial photos of the American Southwest taken from 45,000 foot altitude from Los Angeles to Denver

2CD set – 18 hours - The Physics of Ancient Celestial Disasters

OTHER PRODUCTS

Emergency Survival Re-Calendar Portable Sun-Star Clock

3 Poster Set – Historically important Poster Set

Professor McCanney hosts a weekly radio show on station WWCR – Nashville, TN—"The James McCanney Science Hour – At the Crossroads"

See www.jmccsci.com for details

Made in the USA
Monee, IL
04 November 2026

67383328R00105